Y0-BRS-122

WITHDRAWN
LIBRARY
College of St. Scholastica
Duluth, Minnesota 55811

ECONOMICS AND TECHNICAL CHANGE

ECONOMICS
AND TECHNICAL
CHANGE

Arnold Heertje

A HALSTED PRESS BOOK

JOHN WILEY & SONS
New York

HD
45
.H3513
1977b

©1973 H. E. Stenfert Kroese B.V.

English translation © 1977 Arnold Heertje

Published in the U.S.A.
by Halsted Press, a division
of John Wiley & Sons, Inc.
New York

ISBN: 0-470-99189-5

Library of Congress Catalog Card No: 77-76757

Printed in Great Britain

Contents

LIBRARY
College of St. Scholastica
Duluth, Minnesota 55811

Foreword

On the occasion of the fiftieth anniversary of the Postal Payments System, the former Director General of the Dutch Post Office asked Professor Heertje to write a book on Economics and Technical Development. His work resulted in a Dutch text, of which the present book is the English translation.

Although himself a mathematical economist, Professor Heertje has chosen to write a basically non-mathematical book on the subject, in which he deals with the history of economic thought on technical development, the history of technology, production theories and the significance of technical development for economic growth, monopoly power and economic policy.

The history of technology is discussed in two chapters, one for the period of up to 1900 and one for this century. The three chapters on production theory deal with the precursors and with microeconomic and macroeconomic production theories. Quite rightly, the author gives a prominent place to the late Erich Schneider's lucid treatment of the subject and adds a description of the diffusion of technical knowledge, so often supposed—in sharp contrast to reality—to be instantaneous.

In the chapter on macroeconomic production theory Professor Heertje pays special attention to the existence of macroeconomic production functions, which is the central issue here. Similarly, in the chapter on technical development and growth, he deals in depth with the distinction between embodied and disembodied technical development.

What seems to me to be of special interest to English-speaking readers are the careful and balanced treatment of the subject and the author's first-hand knowledge of the literature in five languages—a typical 'comparative advantage' of a son of a small country. Although written for the non-mathematical reader, the book is exact in its terminology and clear in its exposition, which should enable it to act as a bridge between the English-speaking world and Continental European thought on this subject.

J. Tinbergen

Preface

Mr Reinoud, the former Director-General of the Dutch Post Office, suggested that I should write a book on the economics of technical development in order to fill a gap in the literature. I accepted this challenge and this book is the result. It has been written as a textbook on the economics of technical development and will therefore be useful to undergraduates, postgraduate students and working economists, although the main aspects of the subject are discussed in sufficient detail to make this book also useful for those doing research in this field.

The terms 'technical change' and 'technical development' are used here more or less synonymously; the only difference is that the former is used when discussing the literature, and the latter to express my personal views, for I feel it to be broader than 'technical change'. However, both terms are employed without any implication of 'progress' in other than a purely technological sense.

I have of course taken into account the literature that has appeared since the publication of the Dutch edition and I have also introduced some new ideas which I have developed since then.

I wish to express my gratitude to Mr Reinoud for giving me this challenging task, to his successor, Mr Leenman, for commissioning the English translation, and to the Dutch Post Office for their substantial assistance.

I am deeply indebted to my teacher, Professor Dr. P. Hennipman of the University of Amsterdam, who constantly encouraged me during the work, read the whole manuscript, and let me draw on his wealth of knowledge. I feel, furthermore, highly honoured that Professor Dr. J. Tinbergen agreed to write a foreword to this book.

I would like to express my thanks to Mr Z. St-Gallay, who translated this book into English. I would also like to thank Dr

S. Lofthouse and Professor Dr N. de Marchi for reading the translation and giving me some useful hints about adapting the book for an international public. However, as the author, I am of course fully responsible for its shortcomings, as well as for any 'positive external effects' it may have. I dedicate this book to my father, whose figure as a creative technical man I had before me while writing, which is why I could put not only my mind but also my heart into the work.

Arnold Heertje

January 1977
Naarden, Holland

1 Introduction

*It is doubtful that we shall ever be able to 'explain'
technical progress to our own satisfaction since
technical progress almost by definition involves the
appearance of the unforeseen*

Dewey[1]

1.1 ECONOMICS, TECHNOLOGY AND TECHNICAL CHANGE

This book deals with those aspects of technology and technical
change which have dominated economic literature. Since we
shall confine our attention to the economic aspects of technical
change, our approach here is somewhat one-sided. This is
reinforced by the fact that we can discuss only a part of the very
large body of economic literature, since technical change affects
all economic phenomena. Consequently some determination of
the limits of the field is necessary. The tension between wants or
needs and the means to satisfy these is directly influenced by
technical change both from a qualitative and a quantitative
point of view. This relationship between scarcity and technical
change has only recently been studied in detail, although
elements of an analysis have always been part of economic
debate. Our analysis deals both with the consequences of
technical change and with the factors that govern it.

The way economists treat this subject is constantly changing,
and this explains the considerable difference of opinion on
several points, such as the causes and effects of technical change
and the role the latter plays, and ought to play, in everyday
life. Optimists point out that better and better durable
consumer goods are available to most people, while pessimists
point out that automation can lead to unemployment and
more or less enforced leisure on a large scale. Again, some
authors stress the prosperity that technical change brings, while
others stress its horrors; some point to the greatly increased
food production made possible by new and better techniques,
while others point to the terrifying wars that modern

sophisticated weapons permit. In addition, there have recently been signs of a more widespread public apprehension about technical change. From the point of view of economic policy-making, technical change is regarded by some as an inevitable and uncontrollable process and by others as a process that can be planned and regulated. Half the world thinks that science and technology flourish best in freedom, while the other half thinks that only central planning and control can ensure full growth.

I have not outlined by any means the full scope of the controversies, for there will be ample opportunity later on to deal with the disputes found in the literature. However, certain aspects and consequences of technical change have been so widely discussed since economics became a more or less independent branch of science that a preliminary account of them should be given straightaway.

1.2 SOME OF THE MAIN PROBLEMS

With the advent of the Classical economists, ideas about the significance for production of the factors of production (land, labour and capital) came to the forefront of economic debate, and economists gradually developed the notion of a quantitative connection between the level of production or output and the quantities of the factors of production. This work has recently culminated in a sophisticated mathematical and economic analysis of the production function and in the econometric verification of this relationship, but the study of the evolution of this concept from its beginnings is still worthwhile. The Classical economists' ideas about the effect of diminishing returns on economic growth gave a more precise form to initially vague observations about social phenomena, and are still in many respects valuable today—though without the mass of details that often confused the Classical economists themselves and distracted them from the important issues. The way the Classical laws of returns led to the modern, refined theory of production will be examined in this book, because the connection between production and the factors of production is an important technical relationship.

The production function relates the level of output to the level of input of the factors of production at a given state of technical knowledge. An increase in technical knowledge causes a shift in the production function, so that technical change is often identified with a shift in this function. Since the range of technical possibilities can increase even without an increase in the stock of technical knowledge, for example by education and the diffusion of existing technical knowledge, we must make a clear distinction between technical change in the broad sense—meaning the *development* of technical possibilities—and technical change in the narrow sense—meaning additions to the stock of technical knowledge. If the production function is viewed as an expression of the technical possibilities, then technical change in the broad sense corresponds to a shift in the production function, and—conversely—such a shift indicates technical change in the broad sense. There is then no need to talk about technical change in the narrow sense, though this also implies a shift in the production function, since increased technical knowledge creates new technical possibilities. But one of the questions that must be answered is whether the production function can be regarded both as a microeconomic and as a macroeconomic entity.

The application of new technical possibilities is often included in the description of technical development, and this interpretation is in fact the dominant one in everyday language. There is a connection between the application of techniques and the creation of new technical possibilities if such an application leads to new technical knowledge. Although the application of new techniques as such will not be treated here as technical change, it is an aspect that will receive as much attention as the increase in the range of technical possibilities. However, we cannot really discuss the difficult problem of terminology and definitions, until we have made a more precise definition of the concepts involved.

There have been sharp exchanges between economists about whether the application of new techniques ousts labour from the production process: some say 'no', for the people who thus lose their jobs soon find new ones, while others say 'yes', for the capital stock does not increase fast enough to create new jobs for all those who have become redundant through the introduction of labour-saving technology. Technological unemployment is a recurrent—and currently very topical—theme in economic literature.

When examining the way technical change alters the relative power of the producers and the owners of the factors of production, one inevitably thinks of Marx (irrespective of whether one sees him chiefly as a philosopher, a sociologist, an economist or a social reformer), and his contribution to the subject will be examined at some length. Disentangling the interactions between changes in technology and changes in the type of market competition is another important issue. The discussion of this point is not only of theoretical importance, it is also characterized by specific political considerations, owing to the irreversibility of the social changes involved. Care is needed here to differentiate between statements that involve a value judgment and those which do not.

The role of technical change in economic growth should be mentioned separately, though it is also involved in the main problem, that is, the general relationship between production and the factors of production. This separate mention is justified partly because great stress is placed on the dynamic interpretation of the production function in growth theory, and partly because so much of the literature deals with the contribution of technical change to economic growth.

This brief outline of some of the main problems should not be taken as an indication of a strictly compartmentalized treatment of these topics, and in fact the boundaries between them will often be ignored for the sake of clarity and the presentation of the argument; this shifting focus itself reflects the uncertainties and turmoil that are the by-products of rapid and often spectacular technical development.

1.3 BRIEF SUMMARY OF THE CONTENTS

Since one of the main aims of this book is to trace the history of the relevant economic theories, the approach chosen here is usually chronological. This will illustrate the diversity of the treatments given to the recurring main topics, and gives a more profound insight into them. Another advantage of this approach is that the relationships—often involved and ramified—will not claim our attention all at once, but will instead move into our field of vision gradually. However, the chronological approach will sometimes be relaxed, notably in the analysis of the work of past authors by modern methods—this being a good way of evaluating to what extent the analysis of the past is still useful—and in the use of new concepts to describe old ideas, whenever this helps to clarify the argument.

Chapters 2 and 3, dealing with the Classical school and Marx, respectively, will set the tone of the whole book, to some extent, by presenting two diametrically opposite pictures: one of the harmony and spontaneous order which the Classical economists saw in economic affairs, and one of the disharmony and concentration of power, with its disintegrating influence, which Marx believed to be the central characteristics of capitalist economies.

Chapter 4 deals with technical developments up to about 1900. Chapter 5 shows that although Marx's view of economic and social development as something closely related to the state of technology was abandoned around 1870 by the subjectivistic economists who set the trend in the period 1870–1910, there was still a strong objectivistic undercurrent in this period—a fact that indicates a certain continuity of development and blurs a sharp distinction between the objectivistic and the subjectivistic economists over price theory.

A whole chapter (Chapter 6) is devoted to Schumpeter. The important position accorded to his contribution is justified by his unconventional and penetrating analysis of the dynamics of capitalism. This is followed in Chapter 7 by an account of technical developments in the first half of this century.

When discussing modern economists, who tend to express their ideas in a mathematical way, for example in the mathematical formulation of the production function, we shall bring in at appropriate points both the schools of thought mentioned above, which are noticeable for example in the marginal utility school, and the re-appearance of Classical ideas, refashioned mainly by von Thünen and Wicksell.

Chapter 8 and 9 deal with modern production theory and the classification of technical change. This is followed in Chapter 10 by a discussion of the theory of growth, where we shall also analyze the results of econometric investigations into the significance of technical change for economic growth. The subject of Chapter 11 is the relationship between technical change and the different types of market competition.

Of the last two chapters (Chapters 12 and 13), the first deals with technical change and economic policy and with some welfare aspects of the former, while the second deals with the changing frontiers of theoretical economics, with economics and social policy, and with the possibilities that lie before mankind for the effective management of invention and innovation.

2 The Classical Economists

It appears that Society, when once placed in a position removed a certain degree above utter barbarism, has a tendency, so far as wars, unwise institutions, imperfect and oppressive laws, and other such obstacles, do not interfere, to advance in Wealth and in the Arts which pertain to human life and enjoyment.

R. Whately[1]

2.1 JAMES STEUART AND ADAM SMITH

James Steuart's *An Inquiry into the Principles of Political Economy*,[2] published in 1767, has been overshadowed by Adam Smith's *The Wealth of Nations*.[3] But Steuart in fact gives a clearer treatment of the recurrent question of whether the introduction of machines is 'prejudicial to the Interest of a State, or hurtful to Population'. His answer is broadly similar to that of the Classical school, the exponents of which put their trust in a rapid return to equilibrium after a disturbance in the system. In the case of sudden mechanization, however, Steuart suggests that the government should take corrective measures;[4] thus, although there are some disadvantages of mechanization, these are only temporary. According to him, experience shows that on the whole the acquisition of more machines brings about improvements. The introduction of machines does lead to temporary unemployment, but their manufacture offers long-term employment and leads to price reductions.[5]

Adam Smith dealt with the introduction of new machines as part of the wider problem of the division of labour—a topic of central interest to him.[6] His contention is that the increase in the productivity of labour is mainly due to the division of labour, and that this productivity rises faster in industry than in agriculture—the only two sectors he recognizes—because the division of labour can be taken further in industry. One of the factors that raise per capita production is 'the invention of a great number of machines which facilitate and abridge labour,

and enable one man to do the work of many', and the invention
of machines 'seems to have been originally owing to the division
of labour'. The more one succeeds in concentrating on a certain
part of the production process, the more likely one is to
discover a way of simplifying this process. He added that, for
this reason, many machines are developed by 'common
workmen', whereas numerous other improvements in machines
are brought about not by their users, but by the 'ingenuity of
the makers of the machines'.[7]

Discussing mercantilism, Smith challenged the prohibition
on exporting machines and argued in favour of the free exchange
of capital goods—and of the knowledge embodied in them—
with the motto 'consumption is the sole end and purpose of all
production'.[8] It may also be pointed out, as Hollander[9] has
done, that Smith did not make a sharp distinction between
technical change, on the one hand, and internal and external
economies of scale, on the other. Nor did he distinguish
carefully between the generation of new knowledge and the
actual application of new techniques. On the other hand, he
shows a remarkable insight by recognizing both labour-saving
and capital-saving technical change. Hollander stresses that
Smith in fact dealt with many technical developments prior to
1776, though not those which had occurred in the cotton
industry. G. B. Richardson has even expressed the view that
'technological progress for Smith is not an extraneous
circumstance affecting economic growth but integral to his
theory of economic development'.[10]

Steuart compares favourably with Smith in that[11] he dealt
with the effect of new machinery on economic life more
systematically.[12] However, they both discussed problems that
are still of interest today, such as the more extensive use of
machines, the conditions under which inventions are made, the
role of the makers and users of machines, unemployment, the
effect on productivity, and the increase in production.

2.2 EDWARD WEST

A pamphlet[13] on the investment of capital in agriculture was
published anonymously by Edward West, a lawyer, in 1815.
This essay, which is incidentally still worth reading today, deals
with the principle that 'in the progress of the improvement of

cultivation the raising of crude produce becomes progressively more expensive'. West thus paved the way for the formulation of the law of increasing and diminishing returns—one of the most important technical relationships in economics.

In connection with Smith's statement that new machines and the division of labour raise productivity in agriculture less than they do in industry, West said that these factors are responsible for only a small part of the difference between the two sectors, because 'the effects of the subdivision of labour and the application of machinery are considerable even in agriculture', and that the really important point, overlooked by Smith, is that 'each equal additional quantity of work bestowed on agriculture yields an actually diminished return'. The concept of diminishing returns was thus introduced into the argument, although it is not clear whether West was thinking here of average or marginal returns.

West distinguishes between an intensive and an extensive variant of this law; in the first case, the returns diminish because of the increasingly intensive cultivation of the same land, while in the second case they diminish as less and less fertile land is cultivated with the same amount of labour and capital as used on fertile land. As the population grows, people are compelled to cultivate the less fertile land, because the returns on the more fertile land have diminished.

West did not see any further consequence arising from this recognition of the two causes of diminishing returns.[14] The production function linking output with the factors of production had not yet been formulated. Today we would say that West did not distinguish between a movement along the graph of the production function and a shift of the graph itself. But West's treatment is not so abstract, for he simply wanted to analyse the physical conditions of economic production apropos of the Corn Laws then being debated at Westminster. In the meantime, however, Malthus and Ricardo were writing their own pamphlets on this subject.

2.3 MALTHUS AND RICARDO

West's pamphlet was published on 13 February 1815; Malthus's pamphlet had appeared ten days before, and Ricardo's *Essay on Profits* about ten days after.[15] However, it would be pointless to look for a precise analysis of diminishing returns in these works of Malthus and Ricardo—such dissimilar masters of the Classical school. Their formulation of the problem is less accurate than West's, although there are many signs that they too based their argument on the decrease in agricultural productivity that occurred when less and less fertile land was brought under cultivation. Malthus's biographer, Bonar, acknowledges that West's treatment of the problem is simpler and clearer than Malthus's.[16]

According to Malthus, land is a set of machines, which is steadily improved, but whose productivity falls behind that of industry as the latter is mechanized, since these 'machines' are merely 'gifts of nature'. The stress Malthus puts on this productivity differential suggests that he assumed different rates of technical development for agriculture and industry. He goes on to point out that it is possible to produce enough capital-saving and labour-saving machines in industry, so that the price of goods is 'reduced to the price of production from the best machinery',[17] whereas even the most fertile land cannot keep up with the 'effective demand of an increasing population', and so while industrial products become cheaper, the price of agricultural produce rises 'till it becomes sufficiently high to pay the cost of raising it with inferior machines and by a more expensive process'. The expression 'inferior machines' is used here in place of 'less fertile land' (the extensive variant of diminishing returns).

Ricardo had a vague notion about diminishing returns even before 1815,[18] and he must have had some idea of their effect on profits and rent, for it took him only a few days to write his *Essay*.[19] He did not know about West's pamphlet at that time but West must have introduced himself to Ricardo shortly after the *Essay* had been published, for on 9 March 1815 Ricardo wrote to Malthus that West agreed with his views. Only then did Ricardo read West's book, finding himself to a large extent in agreement with it.[20] Ricardo's *Essay* contains, besides a description of the extensive variant of diminishing returns, some other passages on mechanization that are of interest, though they have not been widely discussed. Ricardo puts the invention of machines on a par with the expansion of international trade and the division of labour, on the grounds that all three increase the amount of goods and 'contribute very much to the ease and happiness of mankind'. His concern for the common good is again evident in his argument against Malthus' advocation of the restriction of grain imports, because of possible losses of investment in agriculture, when he says that, by the same token, it was wrong to introduce the steam-engine and improve Arkwright's spinning frame, on the grounds that 'the value of the old clumsy machinery would be lost to us'. Reflecting on Malthus' likening land to machinery, Ricardo asked what the sense was in using the worst of these machines when 'at a less expense we could hire the very best from our neighbours'. He also thought that in 1815 employees would only benefit from improved machinery, which, 'it is no longer questioned, has a decided tendency to raise the real wages of labour'. He concluded that impeding the import of cheap grain was like resisting technical improvements: 'To be consistent then, let us by the same act arrest improvement, and prohibit importation'.[21]

2.4 THE 'PRINCIPLES' OF RICARDO AND MALTHUS

Ricardo had to be urged by James Mill[22] to write his *Principles*,[23] for he felt much hampered by 'want of talent for composition'.[24] It is true that in all three editions published in his lifetime, the *Principles* does read more like a collection of essays than a single coherent work.

Diminishing returns are discussed mainly in Chapter 2, 'On Rent', where they are said to arise mostly from less fertile land being brought under cultivation; however, Ricardo was also aware of the intensive variant of the law of diminishing returns when he said 'capital can be employed more *productively* on those lands which are already in cultivation'.[25]

As for the economic significance of mechanization, the remarks made in Chapter 1 show that Ricardo was mostly concerned with the price-reducing effect of labour-saving machines, which he found compatible with rising wages, the introduction of machines being the consequence of labour becoming more expensive. The longer the useful life of the machines, the lower the prices would drop. It is important that Ricardo's view that mechanization would benefit the whole community refers to the accumulation of capital in the course of time, the ratio between fixed capital and circulating capital being determined by the relation between the wage rate and the return on capital.

In reaction to Ricardo's ideas, Barton[26] voiced some pessimistic views about the effect of mechanization on the position of the working classes. He argued in particular that the increasing demand for manpower depends on the changing ratio between fixed and circulating capital. Ricardo echoes this and states expressly that 'in proportion as the accumulations of capital are realized in fixed capital, such as machinery, buildings, etc. they will give less permanent employment to labour, and therefore there will be a less demand for men, and less necessity for an increase of population, than if the accumulated capital had been employed as circulating capital'.[27] This again indicates that what Ricardo had in mind was the process of the accumulation in the course of time.

The significance for the demand for labour of the ratio between fixed and circulating capital can, however, also be illustrated by postulating a given total capital; an increase in the share of the fixed capital in this total can lead to redundancies in certain circumstances. This is perhaps what McCulloch meant when, in his favourable assessment of Barton's publication in the *Edinburgh Review* in January 1820, he said that 'the fixed capital invested in a machine must always displace a considerably greater quantity of circulating capital, for otherwise there could be no motive to its erection; and hence its first effect is to sink, rather than increase, the rate of wages'.[28] Ricardo did not realize that what McCulloch had in mind here was technical change embodied in new capital goods, for he wrote to McCulloch that 'The employment of machinery I think never diminishes the demand for labour—it is never a cause of a fall in the price of labour, but the effect of its rise.'[29] Ricardo continued to regard mechanization as the outcome of the substitution of capital for labour under the influence of higher wages. From the correspondence between these two authors during 1821 it transpires that Ricardo managed to convince McCulloch, which proves that the latter had also no clear idea of the various possible interpretations of a change in the ratio between the fixed and the circulating capital.

In April 1820, a year after a second and rather better

constructed edition of Ricardo's *Principles* had appeared, Malthus published his *Principles*.[30] These two authors had known each other for nearly ten years, but this did not prevent Ricardo from saying that Malthus thinks 'he may use words in a vague way, sometimes attaching one meaning to them, sometimes another and quite different',[31] and this criticism is not entirely unfounded.

Malthus believes that Barton is right in some instances when he says that 'the demand for labour can only be in proportion to the increase of circulating, not of fixed capital', but Malthus generally believes that mechanization can reduce prices, which enlarges the market and this in turn makes for a 'great demand for labour'. Time and time again Malthus stresses the need for an 'adequate market', without which the labour-saving inventions are less beneficial. However, he also points out the military significance of new inventions: 'In carrying on the late war, we were powerfully assisted by our steam-engines'. All in all, Malthus believes that production can increase through the accumulation of capital, the fertility of land and the use of labour-saving inventions. Whether these possibilities are utilized depends on the 'continued increase of the demand for commodities'.[32]

In the autumn of 1820, Ricardo made some casual notes on Malthus's book, which are sometimes contradictory. For example, about Malthus's comments on Barton, Ricardo says 'If capital is realized in machinery, there will be little demand for an *increased*[33] quantity of labour', and a little later he says that 'with a cheaper mode of cultivation the demand for labour *might* diminish'. This contradiction disappears if a distinction is made between the accumulation of capital in the course of time and the change in the composition of a given total capital.

While Malthus sees disadvantages to employees from mechanization, especially if the market for the products is not large enough, Ricardo believes that 'the individual may suffer but the community benefits'. He is again mainly thinking here of the falling prices of goods. Ricardo says Malthus exaggerates the advantages of the steam-engine in war, and concludes that it is difficult to find out what Malthus's views are on the introduction of machines. When the world can be regarded as a single large country, Ricardo and Malthus agree that 'the most extensive use of machinery' is a desirable thing, but when one region has no dealings with the rest, Ricardo still believes that one is faced with 'unmixed advantages from the accumulation of capital, improved fertility of the soil and invention to save labour'[34]—since the demand will increase as well—while Malthus has considerably less faith in the rise in demand resulting from increased production. This shows a different interpretation of Say's law; but before discussing this topic we must examine Ricardo's views on mechanization in the third edition of his *Principles*, published in 1821.

2.5 RICARDO'S REVISED VIEWS

In the chapter on machinery, Ricardo himself admits that he has considerably changed his views on the effect of machinery on industrialists, landowners and workers. This is the basis of the conventional opinion that Ricardo's revised, pessimistic appraisal of the effect on the position of labour is incompatible with his original view, described in Section 2.4 above. The present author believes, however, that Ricardo's new analysis complements his earlier published view and that Ricardo's revised position only represents a change from an unpublished view. Indeed Ricardo himself says that 'I-am not aware that I have ever published any thing respecting machinery which it is necessary for me to retract'.

To understand Ricardo's thought, we must realize that we are not dealing here with mechanization at a given state of technology, but with that technical change which is embodied in new machines. Furthermore, we must clearly distinguish the production of capital goods from the introduction of machines. Originally Ricardo's starting point was that like other social groups the workers would profit from mechanization because of a fall in the price of goods. Now, however, he comes to the conclusion that the replacement of labour by machines is 'often very injurious to the interest of the class of labourers'.[35] He adds that his mistake came from thinking that an increase in total profits would be accompanied by an increase in the wages.

Ricardo's argument can best be illustrated by examining the example he gives:[36] a capitalist has a capital of £20,000, of which £7,000 is fixed capital (buildings, etc.), and the rest is circulating capital, used to pay the labour force. His annual profit is 10 per cent, i.e. £2,000. Like the circulating capital, this profit can be imagined to be in the form of consumer goods. At the beginning of each year, the capitalist has £13,000, from which he pays the wages. At the end of each year, he has £15,000, so that he can again spend £2,000 on consumer goods in the next year. Let us now assume that one year he instructs half his labour force to build a machine, while the other half carry on producing consumer goods. His wage fund is again £13,000. In this year his factory produces £7,500 worth of consumer goods and a machine also worth £7,500. His profit is again £2,000. He now dismisses the workers who made the machine and puts the latter into production instead. His wage fund for the next year has now dropped from £13,000 to £5,500. The machine and the remaining labour force now need only to produce goods worth £7,500 to cover wages and profit. The whole example is illustrated by Figure 1.

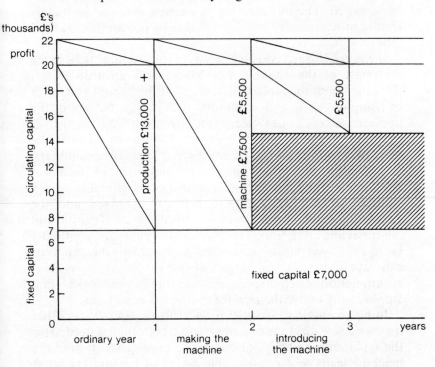

Figure 1

The question may arise at this point what advantages Ricardo's capitalist has from making and introducing a machine, since his profit is still the same, i.e. £2,000 per annum and his total capital is still only £20,000. Ricardo's basic assumption is that mechanization lowers the price of consumer goods, which implies that the useful life of the machine is a number of years. The increase in the net income then applies also to the real income of the capitalist, while the drop in the wage fund reflects the dismissal of half his labour force. The purchasing power of the workers who are still employed by him has also increased. It appears from this case that Ricardo assumes a constant money wage. If the capitalist halved the wages, he could re-employ the other half of his labour force. Finally, it is an important assumption underlying Ricardo's argument that the original fixed capital remains the same. The extent of his premises and buildings remains the same, though the labour force has been halved. The introduction of the new machine seemingly results in a saving of only one factor of production, namely labour.

It emerges from a careful analysis of this example that, in the first part of the chapter 'On Machinery', Ricardo is not thinking about the replacement of labour by capital as a result of rising wages at a given state of technology, but about the application of a new technique or technology which is embodied in a machine and is expected to lead to a drop in the price of consumer goods. He concentrates here on a shift in the composition of capital under the influence of technical development, and not on the accumulation of capital at a given state of technology, the ratio between the fixed and the circulating capital being dependent on the price of capital and labour at any given time. The dismissed workers can eventually be re-employed, since—according to Ricardo—the capitalist can save more, for his outgoings remain the same. As this accumulation of capital proceeds, more and more workers find employment under the new technological conditions.

In his recent analysis of Ricardo's theory, Hicks[37] says that making a new machine requires more labour than is needed for the replacement of the old one, while the introduction of the machine leads to a considerable saving of labour. The wage fund at first decreases, but then increases again when the capitalist invests his extra profit, so that the displaced workers can be re-employed. This interpretation is certainly related to Ricardo's, but the dynamic context used by Hicks is so different from that of Ricardo's that it may easily lend support to the misunderstanding that Ricardo fundamentally changed his mind about the effect of mechanization on the level of employment.

The presentation of his chapter 'On Machinery' clearly indicates that we are confronted here with a supplement to his analysis. By a horizontal line Ricardo separates the discussion of the change in the composition of capital from the rest of the chapter and starts the second part with these words: 'Independently of the consideration of the discovery and use of machinery, to which our attention has been just directed, the labouring class has no small interest in the manner in which the income of the country is expended'. This second part also deals with the accumulation of capital in the course of time, the effect of rising wages on the composition of the total capital, and the resulting demand, due to population growth, for more production. This discussion, therefore, does not touch on technical change, a topic that is central to the argument in the first part of the chapter. Since Ricardo runs these two parts together in one chapter, it is not surprising that analysis of them has been lumped together. Ricardo remarks:

The statements which I have made will not, I hope, lead to the inference that machinery should not be encouraged. To elucidate the principle, I have been supposing that improved machinery is suddenly discovered, and extensively used; but the truth is, that these discoveries are gradual, and rather operate in determining the employment of the capital which is saved and accumulated, than in diverting capital from its actual employment.

This passage shows that Ricardo distinguishes not only between technical change and mechanization, but also between (a) a sudden change in the composition of a given total capital and (b) the distribution of the new capital between fixed and circulating capital. The following quotation shows that no connection should be sought between Ricardo's revised thoughts on mechanization and his famous statement about the constant competition between labour and capital; this statement fits in with his theories of the accumulation of capital, propounded earlier:

With every increase of capital and population, food will generally rise, on account of its being more difficult to produce. The consequence of a rise of food will be a rise of wages, and every rise of wages will have a tendency to determine the saved capital in a greater proportion than before to the employment of machinery. Machinery and labour are in constant competition, and the former can frequently not be employed until labour rises.[38]

Ricardo is undoubtedly thinking here of the substitution of capital for labour under the influence of price changes at a given state of technology—the connection between mechanization and wage increases which is now known as the 'Ricardo effect'.[39] What Ricardo says is that the demand for labour increases with increasing capital, but not in proportion with the latter. He quotes Barton's pamphlet of 1817 with approval in this connection. This indicates that his later thoughts on mechanization were not a delayed reaction to his discussion with Barton in 1817, but arose instead quite independently of these.[40] His agreement with Barton's views is not at variance with his earlier opinion.

Apart from pointing out that the improvement of machines is a gradual process,[41] Ricardo also maintains that if the government curbed mechanization, the capital would migrate elsewhere, and 'this must be a much more serious discouragement to the demand for labour, than the most extensive employment of machinery'. When part of the capital is invested in 'improved machinery, there will be a diminution in the progressive demand for labour; by exporting it to another country, the demand will be wholly annihilated'. The use of better machines can lower the production cost, and the goods can be sold abroad at a lower price; conversely, if mechanization is discouraged while it is encouraged in other countries, one is forced to 'export money, in exchange for foreign goods'[42]—with an unfavourable effect on the balance of trade.

Ricardo himself sometimes gives the impression that his example of technical change[43] embodied in new capital goods conflicts with his original theory insofar as the effects of mechanization on the demand for labour are concerned, but the relevant parts of his system in fact complement one another. The new discussion of mechanization first appearing in the third edition of his *Principles* cannot be compared with the mainstream of his thought insofar as he did not work out the long-term effects of the case he gave as an example. This leaves some questions unanswered about the lifespan of the new machines [44] and about the hypothesis that the real profit of the capitalist remains unaffected by the introduction of machines. The main point for Ricardo is that when labour-saving machines are installed, workers are withdrawn from consumer-goods production, so that the wage fund, i.e. the circulating capital, decreases. With a given total capital, the fixed capital increases, so that the demand for labour, which depends on the amount of the wage fund, is reduced, as the wages remain the same. The resulting unemployment must be distinguished from the unemployment which Malthus and especially Sismondi [45] thought would arise from demand falling behind production as the latter rises with mechanization. McCulloch, who is utterly confused by Ricardo's allegedly modified views on mechanization, treats Ricardo's opinion as a very special case in his book published in 1825,[46] while dissociating himself from Barton.[47]

Ricardo's views on the change in the ratio between fixed and circulating capital appears in a completely different light in the context of the history of economic thought if we replace the words 'fixed' and 'circulating' by Marx's terms 'constant' and 'variable'. Variable capital is invested in labour and it 'reproduces the equivalent of its own value, and also produces an excess, a surplus value'.[48] 'The instrument of labour, when it takes the form of a machine, immediately becomes a competitor of the workman himself. The self-expansion of capital by means of machinery is thenceforward directly proportional to the number of the workers, whose means of livelihood have been destroyed by that machinery.' With each improvement in the machinery, fewer people will find employment. The ratio between the constant and the variable capital—insofar as it is determined by the technical composition of the capital—is called the organic composition of capital. As capital is being accumulated, the variable component decreases, while the constant component increases. 'The labouring population therefore produces, along with the accumulation of capital produced by it, the means by which it itself is made relatively superfluous . . . to an always increasing extent'.[49]

These quotations show that Marx's theory about the increasing share of constant capital in the organic composition of capital (and about its effect on the demand for labour) can be regarded as a dynamic and 'absolutized' version of Ricardo's treatment of technical change. We shall see in Chapter 3, when dealing with Marx, how far technological unemployment, recognized by Ricardo, can be reconciled with Say's law.

2.6 SAY'S LAW

Since the beginning of the last century, Say's law has been interpreted in very different ways.[50] It is sometimes taken to mean—in particular by James Mill[51]—that production and demand must necessarily be equal. This view is opposed by all those who assume that all the means or factors of production are employed in the long run, conceding the validity of Say's law only for the long run.[52] However, it appears from the original formulation of his law that Say thought it applied in the short run as well.[53]

The best way of discussing the Classical economists' view of Say's law is to distinguish between an elementary and a sophisticated formulation of the law. For this purpose, a closed economy without a public sector is assumed, and the following variables are used: the national product W, the national income Y, and the total expenditure B, i.e. the sum of investment and consumer expenditure. Say's law as such does not state that all the factors of production will be employed, but in fact such a situation, brought about by the flexible operation of the price mechanism, was a separate postulate of Classical thinking. Therefore, in describing the two formulations of Say's law, national product, W, is taken to have its maximum possible value W^* (i.e. $W = W^*$). It follows from this and from $Y = W$ (which means that the national income is equal to the value of production) that $\bar{Y} = W^*$, where $\bar{Y}$ is the equilibrium value of the national income. The condition for equilibrium is that the total expenditure $\bar{B}$ should be equal to the national income $\bar{Y}$, which gives Say's law in its elementary formulation, $\bar{B} = W^*$.

This says that at equilibrium, maximum production is exactly equal to expenditure, the money involved in the latter being made available—in the form of income—by production. We call this interpretation of Say's law elementary, because the statement that the production W^* creates a demand $\bar{B}$ that is quantitatively equal to it is not a behavioural equation, but a necessary consequence of the Classical postulate $W = W^*$, of the identity $Y \equiv W$, and of the equilibrium condition $\bar{B} = \bar{Y}$. This formulation of Say's law explains Say's and Mill's statements, the gist of which is that the law of markets (*loi des débouchés*) is so obvious that no further proof is needed to substantiate it.

The sophisticated formulation of Say's law should also be distinguished from the assumption that all the factors of production are employed, i.e. that $W = W^*$. According to this interpretation, the law states that a given production W creates a demand B of the same magnitude. The equation $B = W$ is now a behavioural equation. It follows from the combination of this result with the Classical postulate that the equilibrium value of the expenditure is equal to the maximum value of production W^*. The equality $\bar{B} = W^*$ is not postulated here but is instead derived—and specifically from the behavioural equation that the expenditure will always be equal to the value of production. This interpretation is consistent with Ricardo's version of the law, according to which no-one produces anything without having its sale or consumption in mind, so that he 'necessarily becomes either the consumer of his own goods, or the purchaser and consumer of the goods of some other person.'[54]

According to the elementary interpretation of Say's law, there can be no general over-production (i.e. no positive value for $W^* - B$). In contrast, the sophisticated version does not rule out such a general disequilibrium, since the actual magnitude of the production W can differ from the maximum value W^*. If we combine the sophisticated version with the Classical postulate that all the factors of production are in operation, general over-production is impossible, since the expenditure at equilibrium is adjusted to the maximum production.

Partial disequilibria are possible in both versions, since the equality of production and expenditure is compatible with a difference between the composition of production and the composition of demand. The price mechanism removes such partial disequilibria, but this does not mean that Say's law should be identified with a flexible price mechanism which re-establishes an equilibrium.[55]

Malthus,[56] among others, disputed the validity of the statement that general over-production is impossible. He believed that when capital accumulation produces an exceptionally large quantity of goods and if the landowners' and the manufacturers' purchasing power and propensity to buy are assumed to drop, the price of goods also drops, so that the profit vanishes and production stagnates. One is then faced with over-production, which 'in this case is evidently general, not partial'. This is nothing but a straightforward denial of the validity of Say's law in its sophisticated formulation.

In one of his *Essays*, written at about the age of twenty-four, John Stuart Mill analyzed Say's law in depth.[57] The gist of what he says is that if money is regarded as a good, no general over-production of goods can arise, but otherwise a general over-production is just as feasible as a partial over-production, even if both are exclusively temporary phenomena. The main point is that a permanent disequilibrium is impossible, but a partial or a (temporary) general one may result from a 'want of commercial confidence'. Mill's special contribution lies in a less rigorous concept of Say's law, according to which the expenditure B_t need no longer be equal to the production W_t in any given period t. He believes that in the long run the total demand for goods is determined by their total supply. Short-term general unemployment is possible and thus not incompatible with his version of Say's law. Mill's view can in fact be regarded as a refinement of the sophisticated formulation of Say's law, according to which expenditure at any time is determined by production.

To sum up, the elementary formulation of Say's law is tantamount to the formal equilibrium condition $\bar{B} = W^*$. When an attempt is made to put this into words, it is easy to give the impression that the expenditure necessarily arises from the production, and this is in fact the impression one gets when reading, in particular, the relevant passages by James Mill and certain parts of Say's explanation. In fact what Say meant, and what Ricardo said, corresponds more to the sophisticated form of the law, according to which the expenditure B is always equal to the value of production W—a behavioural equation. John Stuart Mill's interpretation is a variant of this form of Say's law, because he does not suppose that the supply is in each period balanced by an equal demand. Mill recognizes the possibility not only of a partial disequilibrium, but also of a transient general disequilibrium, in the goods sector, so that for him the law is valid only in the long term.[58]

2.7 SAY'S LAW AND TECHNICAL CHANGE

Both excessively fast and excessively slow technical change will undermine the equality between expenditure and production postulated by Say's law. Too fast a mechanization can lead to a supply of consumer goods that demand cannot immediately absorb. This happens particularly when the employees who have been made redundant cannot be immediately re-employed, for example because of insufficient complementary capital goods. On the other hand, too slow a technical change can lead to a situation in which selling in certain sectors becomes sluggish owing to relatively high prices. This possibility is mentioned by Say himself in a chapter in his *Cours* that deals with the 'limits of production'. Here he speaks about a stage where the production ceases to increase and the price of the goods exceeds their utility, partly because 'the production techniques are not sufficiently advanced to permit production at a low cost'.[59]. This lag in technical development undermines Say's law in the strict sense; furthermore, some unemployment can also be generally expected.

Unemployment resulting from the introduction of machines had already been mentioned by Say in the first edition of his *Traité*, where he says that, when workers lose their jobs as machines are installed, their position is 'truly deplorable' unless there are some additional capital goods ensuring their re-employment. The disadvantages are temporary and incidental when the new machines are invented in a society where capital formation is taking place. It is interesting that Say points out that new machines do not only save labour, but also change the nature of the goods, so that 'rejecting a new machine means rejecting a new product'.[60]

So far, the gist of Say's argument is that if technical change is so fast that capital formation cannot keep pace and cannot provide new jobs for the dismissed workers, unemployment may arise. Say's law then ceases to hold, with expenditure lagging behind rising production. However, Say does not think it very probable that this situation will arise, partly because mechanization is a slow process and partly because the higher the degree of mechanization, the more difficult it is to invent new machines. Besides, he says that entrepreneurs tend to wait and see which way the wind blows.[61]

If unemployment still occurs, forces which re-establish the original situation appear. This is described in the second edition of the *Traité* in the following terms: mechanization in a certain sector creates such a large supply of goods that prices fall; this brings the goods within more people's reach, and this increased demand in turn creates new jobs.[62] In the fourth edition of the *Traité*, however, this equilibrium mechanism recedes into the background, and Say examines more closely the employment changes produced by new machines in the various sectors of the economy. Say also speaks of mechanization creating new jobs (compensatory employment).

It can thus be seen that Say recognizes compensating factors both on the demand and on the supply side so that, while too slow or too fast technical change disturbs the equilibrium and jeopardizes the validity of his law, such disequilibrium is not very likely to come about in practice, and if it does, these compensating factors which provide work for the displaced labour force come into operation, and expenditure can again keep up with rising production.

The next question is how far technological unemployment, discussed in Section 2.5, can be reconciled with Say's law. As mentioned before, Ricardo assumes that (a) the new machine is made by workers who have been taken out of the production of consumer goods, and (b) the installation of the new machine brings about a considerable saving of labour. For a certain period, not enough machines are being manufactured to absorb all the displaced labour force. This change on the supply side does not invalidate Say's law in its sophisticated formulation, according to which the expenditure is always equal to the value of production. In Ricardo's view, both the wage fund and the level of production drop (while the profits remain the same), so that the reduced expenditure is equal to the reduced value of production.

Malthus maintains that mechanization invalidates Say's law if it 'adds so much to the gross produce of the country that the commodities produced cannot be consumed'. In contrast, Ricardo contends that the introduction of new machines 'diminishes the quantity of gross produce, and although the inclination to consume is unlimited, the demand will be diminished due to the want of means of purchasing'.[63] Blaug[64] believes that Ricardo exaggerates the contradiction between the sales and the wage fund, effects caused by the displacement of labour. According to Blaug, Ricardo sets too much store by the sudden introduction of a new technique and does not allow for the operation of the Ricardo effect. This indicates that Blaug does not fully appreciate the significance of the division of Ricardo's chapter 'On Machinery' into two parts. The fact that Say's law remains valid in Ricardo's argument but is in Malthus's view invalidated by mechanization signifies an important difference of opinion between these Classical economists.

2.8 THE COMPENSATION THEORY

The advocates of the compensation theory believe in the operation of mechanisms whereby those who have been displaced from their jobs by technical change are re-employed. In contrast, the advocates of the displacement theory believe that the conditions for compensatory employment are not fulfilled sufficiently, if at all. The following brief discussion will present the major arguments both of the compensation school and the displacement school of thought.

Compensating employment can be created both by endogenous and by exogenous factors. The endogenous factors are inherent in the introduction of a new technique, while the exogenous factors operate independently of the application of new techniques. Compensation in the strict sense occurs when the workers who have been displaced by mechanization are re-engaged in the production process as a result of endogenous factors; when this re-employment is incomplete, we speak of partial compensation.

When compensation takes place under the influence of factors not connected with the introduction of the new technology itself, we speak of compensation in the broad sense. When compensation by endogenous factors is insufficient to bring about the re-employment of all the redundant workers, then capital formation may well bring about full compensation in the broad sense. Kruse[65] believes that new capital formation should be disregarded in discussing the compensation theory. The practical significance of compensation should be assessed on the basis of compensating factors of the endogenous type. There is also a difference between immediate and gradual compensation—some compensating factors making themselves felt only after a certain period has elapsed—and the recognition of this difference introduces a dynamic element into the analysis.

The re-employment of the displaced workers in making the new machines is often mentioned as the first compensating factor, producing at least partial re-employment. As we have seen, Steuart recognizes this compensating factor, for he sees the new jobs arising mainly in the production of capital goods. According to Lederer,[66] however, there can in practice be no immediate compensation, since the workers have not been trained to make machines, particularly more sophisticated ones. Kruse points out, furthermore, that the new machines are made before the redundancies arise. Still more important is his point that any compensation involved here is of the exogenous kind, since additional capital formation is needed for it. If new capital formation is disregarded on methodological grounds, one finds that the creation of new jobs in the one industry must be accompanied by a loss of jobs elsewhere. This compensation argument is therefore unsound for a number of reasons.

An endogenous compensation factor operates in the model of the determination of prices in the case of perfect competition. The introduction of new machines shifts the market supply curve to the right, so that the new equilibrium price is lower than the old one; in this situation, the equilibrium quantity of goods is higher. Advocates of this version of the compensation theory state that the production of this greater quantity needs extra manpower, and this demand for labour partially compensates for the loss of jobs due to mechanization. The implication is either that no additional capital is necessary, or that the additional capital is created from the profits resulting from the introduction of the new technology. This compensation argument will be examined in more detail in Section 2.9.

Another consideration is that with the fall in the price of a good the real income of consumers increases, and this stimulates the demand for other goods, shifting the market demand curves for these other goods to the right. This means that the prices of other goods will go up, as will the equilibrium quantities. On the assumption that this extra production necessitates increased labour, this is another argument supporting the compensation theory, provided of course that enough capital is formed, by the introduction of the new machines, to employ more labour. John Stuart Mill, who expressly calls the spending of the extra income liberated a compensating factor, adds immediately that a demand for goods does not in itself imply a demand for labour, 'unless there is capital to produce them'.[67] When the capital formation is insufficient, the decrease in the prices brought about by technical change and the increase in the real incomes connected with it are not enough to ensure the re-employment of all the workers made redundant. However, Mill believes that innovations are introduced gradually, so that the employees will not suffer as a result.[68]

The extent of compensatory employment also depends on whether fixed proportions of labour and capital are assumed in every branch of production, or whether there is substitution. This is so whether the analysis of compensation is restricted to endogenous factors or extended to the exogenous factor of capital formation. In the case of fixed proportions, the degree of capital accumulation determines the demand for labour, while where there is substitution the amount of labour employed in an equilibrium situation depends on the ratio between the price of labour and that of capital. The significance of the production function for the compensation theory will be discussed later (see Section 10.4). In particular, when the production function is formulated in a dynamic context, it is important to specify exactly which type of technical change one has in mind.

The possibility of substitution between the factors of

production also affects the extent to which a drop in wages can act as a compensating factor. The diminished demand for labour, brought about by mechanization, can depress wages. When some of the capital can be replaced by labour, the displaced workers are re-employed, and the disadvantage of mechanization for the workers lies not in unemployment, but in lower wages. When such a substitution is not possible, on the other hand, only compensation of the endogenous type can be expected, provided new capital is formed from the increased profits. As in the case of the production function, it is necessary here to specify exactly the effect of technical change on the distribution of income between the factors of production (see Section 10.6).

Babbage, whose work will be discussed in Section 4.2, made some remarks about the way compensation operates in practice which deserve to be rescued from oblivion. He used statistics for sixty-five textile mills covering the period 1822–32 to show that as the mechanical looms were ousting hand-looms the number of hand-weavers dropped to a third of their original number, while the number of those operating mechanical looms grew by a factor of five, and an almost four-fold increase in production was accompanied by a twenty per cent increase in the total number of jobs. Babbage adds: 'In considering this increase of employment, it must be admitted that the two thousand persons thrown out of work are not exactly of the same class as those called into employment by the power-looms'.[69] This example shows the need to distinguish between new jobs that require retraining and those which do not. Such a distinction is of course foreign to the harmonious and self-regulating world of the Classical economists.

2.9 MODEL OF COMPENSATION IN CLASSICAL ECONOMICS
Steuart, Malthus and Say all assume that via the price mechanism displaced workers are re-employed, mostly in the same branch of industry. Assuming perfect competition, mechanization shifts the market supply curve to the right, to a lower equilibrium price and a higher equilibrium quantity than existed before the introduction of the new machines.

In perfect competition, the equilibrium price is determined by a market demand function and a market supply function, both of which are here assumed to be linear for the sake of simplicity. If we denote the total amount of supply S, the total amount of demand D, and the price p, we have:

$$S(p) = ap + b \qquad\qquad a > 0 \quad \text{and} \quad b > 0 \qquad (1)$$

$$D(p) = cp + d \qquad\qquad c < 0 \quad \text{and} \quad d > 0 \qquad (2)$$

$$D(p) = S(p) \qquad\qquad\qquad\qquad\qquad\qquad\qquad (3)$$

The equilibrium price $\bar{p}$ is derived from these equations in the usual manner as:

$$\bar{p} = \frac{d - b}{a - c} \qquad (4)$$

The equilibrium price is positive if $d > b$. The equilibrium quantity is given by:

$$\bar{D} = \bar{S} = \frac{ad - bc}{a - c} \qquad (5)$$

This quantity is always positive.

Let us now assume that mechanization increases the supply of a good by a quantity Δb, which means that the existing enterprises offer an extra quantity Δb, at each given price. The new supply function is then written as:

$$S(p) = ap + b + \Delta b \qquad\qquad a > 0, \ b > 0, \ \Delta b > 0 \quad (6)$$

The graph of the market supply function shows a parallel displacement in the (S, p) plane.

If we assume the demand function has not shifted, the new equilibrium price is obtained in the form:

$$\bar{p} = \frac{d - b}{a - c} - \frac{\Delta b}{a - c} \qquad (7)$$

This shows that the new equilibrium price is lower by $\Delta b / (a - c)$. The equilibrium quantity will then be:

$$\bar{D} = \bar{S} = \frac{ad - bc}{a - c} - \frac{c . \Delta b}{a - c} \qquad (8)$$

i.e. it has increased by $-[c . \Delta b / (a - c)]$, which means that the magnitude of the increase in the equilibrium quantity is determined by the mechanization-induced extra supply Δb

(called the 'mechanization effect'), the gradient of the market demand curve c, and gradient of the market supply curve a. It is often assumed that the increase in the equilibrium quantity depends only on the elasticity of the demand, and so one concentrates on the corresponding gradient c in the case of a linear demand function. This is a misunderstanding, as can be seen from Figure 2.

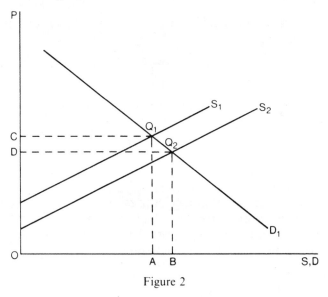

Figure 2

As is customary, the price is plotted here along the ordinate, and the quantity along the abscissa. The shift in the supply curve increases the equilibrium quantity by a magnitude AB. It is now possible to draw another two parallel supply curves through the points Q_1 and Q_2, the gradient of which is different from that of S_1 and S_2, though the increase in the equilibrium quantity has the same value, namely AB. It would follow from this that another increase in the equilibrium quantity arises only if the gradient of the demand curve D_1 is changed. Drawing another set of supply curves through Q_1 and Q_2, different from S_1 and S_2, does not only mean, however, that the gradient is now different, but also that the magnitude of the mechanization effect Δb is different. The increase in the equilibrium quantity by AB in both cases simply means that the combination of a different gradient and a different shift chosen in the second case gives exactly the same result as has been obtained in the first one. However, this in no way detracts from the fact that the mechanization-induced increase in the equilibrium quantity is determined simultaneously by the gradients of the market supply and the market demand curves and by the shift in the supply curve.

Finally, to follow these Classical economists' train of thought as closely as possible, the above case of static comparison must be viewed as a sequence of events, although this entails a certain loss of precision in the analysis. The sequence of events is as follows: new machines are introduced, some of the workers are dismissed, the equilibrium price of the goods produced falls, and the demand for these goods rises. The increase in the demand is determined by the elasticity of the demand curve, provided that one pays attention only to the movement along this curve. However, the new equilibrium point on the demand curve is obtained as a result of a shift in the supply curve. The extra amount demanded is in fact produced, which creates a demand for labour, so that one may speak of partial compensation in this case.

To assess the significance of this compensating factor, one must specify the type of technical change and the production function. The type of technical change will determine which redundancy is accompanied by a shift in the supply curve by Δb. The next important point is whether a new equilibrium state can be achieved with the same amount of capital as the old one, or—if not—whether the extra capital needed can be found from the extra profits resulting from mechanization. The question of whether capital and labour are substitutes for each other also arises, the answer being important in the analysis, for example, of the situation in which the dismissal of workers leads to a drop in wages.

In general then, the Classical economists' analysis leads only to the vague conclusion that some compensation can be expected. To find out whether there will be full compensation a number of assumptions must be made about (a) the type and state of technology, (b) the changes in these and (c) the effect of such changes on the wage rate, the price of goods produced and the demand for capital.

As mentioned before, the compensation theory has been further developed by taking into account the effect of the equilibrium-price decrease on the demand for other goods. A shift to the right in the market demand curves for other goods brings about a rise in the price of these goods, while all the equilibrium quantities increase. The compensation argument here is that a certain labour force is necessary for the higher production in these branches of industry. Whether this is an endogenous compensating factor depends on whether the higher production can be achieved without exogenous capital formation. If capital is withdrawn from other branches of industry, compensation in the branches of industry into which the capital flows is naturally accompanied by redundancy in the former sector. Since compensatory mechanization does not occur in these sectors, it is difficult to see where the capital comes from endogenously, so that endogenous compensation can be expected only if the higher production can be achieved exclusively by the employment of new workers. Again, a detailed analysis in the framework of growth theory is needed to assess precisely the significance of this compensating factor. However, Say's argument can have a second interpretation: his assertion that a drop in prices increases demand may also mean that a lower price increases the overall popularity of a product. The market demand curve then shifts to the right. This argument is that mechanization affects consumer preferences via a drop in price.

To take a closer look at this situation, let us assume that the market demand and the market supply in the above analysis increase in a given period t by Δd and Δb respectively, and that the effect on the demand function makes itself felt immediately. The relationships are then as follows:

$$S_t = ap_t + b + \Delta b.t \qquad\qquad a > 0,\ b > 0,\ \Delta b > 0 \qquad (9)$$

$$D_t = cp_t + d + \Delta d.t \qquad\qquad c < 0,\ d > 0,\ \Delta d > 0 \qquad (10)$$

$$D_t = S_t \qquad\qquad\qquad\qquad\qquad\qquad\qquad\qquad (11)$$

The equilibrium price is then given by:

$$\bar{p}_t = \frac{d-b}{a-c} - \frac{\Delta b.t}{a-c} + \frac{\Delta d.t}{a-c} \qquad (12)$$

It can thus be seen that the equilibrium price is lowered by the mechanization effect Δb and raised by the increased demand due to increased popularity, Δd. If Δb is greater than Δd, the net effect is a drop in the equilibrium price.

The increase in prices, due to the increase in demand, will attract more and more suppliers to this particular market until the increase in the equilibrium price caused by the increased demand is exactly balanced by the drop in the equilibrium price caused by the arrival of extra suppliers, which shifts the supply curve to the right. If the total supply thus rises by γ in each period t, the equations of the model can be written in the following form:

$$S_t = ap_t + b + \Delta b.t + \gamma t \quad a > 0,\ b > 0,\ \Delta b > 0,\ \gamma > 0 \qquad (13)$$

$$D_t = cp_t + d + \Delta d.t \qquad c < 0,\ d > 0,\ \Delta d > 0 \qquad (14)$$

$$S_t = D_t \qquad (15)$$

The equilibrium price is then given by:

$$\bar{p}_t = \frac{d-b}{a-c} - \frac{\Delta b.t}{a-c} + \frac{\Delta d.t}{a-c} - \frac{\gamma t}{a-c} \qquad (16)$$

When Δd is put equal to γ, it follows from Equation (16) that the equilibrium price $\bar{p}_t$ decreases under the influence of mechanization, since the other two effects completely cancel each other out. As Equation (17) shows, the equilibrium quantity will now increase:

$$\bar{S}_t = \frac{ad-bc}{a-c} - \left(\frac{\Delta b+\gamma}{a-c}\right)\cdot ct + \frac{a\gamma t}{a-c} \qquad (17)$$

The last two terms in this equation are positive. The rise in production is determined by the mechanization effect Δb and the extra supply γ produced by the additional suppliers attracted to the market.

Thus new jobs have been created, But again, is this an endogenous compensating factor? The new suppliers must surely need capital, and if this comes from another sector where the goods are not selling well, a net compensation cannot be expected, unless, owing to a difference in the shape of the production functions, the use of a certain amount of capital in sector A is combined with more labour in sector B. If this possibility is discounted, then endogenous compensation is feasible only if the existing firms are assumed to transfer to potential competitors the savings that accrued to them through mechanization.

In conclusion, the endogenous compensating factors discussed in this Section will bring about some compensation, but complete compensation is unlikely. A more accurate result can be obtained only by using the hypotheses about technology and the production process found in the modern theories of production and growth.

2.10 CONCLUSIONS

The Classical economists introduced the law of diminishing returns mainly in connection with the intensive and extensive cultivation of land. In the extensive variant of this law, a given quantity of capital and labour are added to a given quantity of land in each period of time, and as time goes by land of a less and less fertile kind is brought under the plough for cultivation, still with the same amount of labour and capital, so that the returns decrease from one period to the next. The Classical economists obviously could not treat land and capital as interchangeable in this context. As time progresses, the quality of the capital goods either improves or remains the same, but it definitely does not deteriorate.

To arrive at a more general formulation of the law of diminishing returns, we start from the intensive version and assume that units of a variable factor of production of constant quality are combined with a given quantity of another factor of production. Beyond a certain point, the marginal, and hence the average, returns on the variable factor of production decrease. Intensive cultivation of land can be described by a production function that expresses the relationship between the quantity of output and the quantity of inputs, or factors of production, one of these factors being kept constant. It is important that the *quality* of the factors of production is always taken to be the same. The production function in this form reflects the technical possibilities in a given period, and the combination of the factors of production used need not be specified.

The Classical case of the extensive cultivation of land implies a shift in the production function in the course of time, a certain combination of the amount of output and the quantities of the inputs being chosen for each period of the production function.[70] Since each combination of this kind implies a choice which is not discussed in Classical theory, it can be assumed that the Classical economists operated with fixed proportions,[71] in the sense that given a fixed quantity of one factor, they also kept constant the amount of the other factor needed to achieve a certain output, without implying the minimization of total cost.

We have seen that the Classical economists recognized both the technical relationship between the factors of production and the circumstances that influence the nature of this relationship. Here we should mention in particular the effect of mechanization on the volume of production and on the labour/capital ratio, and Say's observation that the popularity of the product itself can be changed by the introduction of machines.[72] The Classical economists' concept of technical change involves for the most part a change in the production function, caused both by qualitative changes in capital and by changes in the end-products. However, since the nature of the changes is usually not defined, the Classical concepts cannot be transformed into a shifting production function with any precision. There are sufficient indications that the Classical economists also recognized the substitution of capital for labour without any new technology. Ricardo's well-known remark that machinery and labour are in constant competition with each other is a recognition of the substitution of capital for labour caused by price changes with a given production function. However, Ricardo's remark does not exclude the possibility that what he meant is a technical change caused by continual wage increases and appearing in the application of inventions embodied in better machines.[73] In that case, it is necessary to distinguish this type of technical change from the introduction of an improved machine that reduces the price of consumer goods and makes some of the labour force redundant.

While these economists may be justifiably said to recognize technical change as an important factor of growth, they tend to believe that it involves only a temporary disturbance in the equilibrium. This is illustrated by Ricardo's remark that the discovery of machines is basically comparable to the expansion of international trade, since both increase the amount of goods at the disposal of the community. Although the Classical economists expressly distinguish between a sudden and a gradual introduction of new machines, they almost always assume that new methods of production are in practice applied gradually.

A gradual technical change is a necessary condition for the validity of Say's law, since it means that the disturbance in the assumed balance between production and expenditure is only temporary, partial and incidental. A rapid technical change may suddenly produce a large gap between the supply of, and the demand for, goods, and make the re-employment of the redundant labour force difficult.

The assumption that technical change is a gradual process in practice is also important for the compensation theory, because the more one can rely on the availability of sufficient capital, the more effective the compensating factors become. According to Kruse's view, with which I agree, compensation in the strict sense exists only if the capital needed is formed as a result of the introduction of a new technology. This view does not exclude, of course, the possibility that in reality the disappearance of technological unemployment is to a large extent due to capital formation occurring independently of technical change, so that compensation in the broad sense takes place.

The rate at which new capital is formed as a result of mechanization also plays an important role in Ricardo's pessimistic conclusions, which refer particularly to the short term, and reveal a lack of faith in the operation of the price mechanism because the redundant workers have no purchasing power. We have seen that in the long run all the unemployed can be re-employed, since the new capital can be formed due to the drop in the price of goods. Ricardo implicitly assumes that unemployment does not reduce wages. In a detailed analysis of Ricardo's view, Wicksell points out that a drop in wages also makes the original methods of production profitable once more and these will absorb the surplus of idle labourers'.[74] We can add to Wicksell's criticism of Ricardo the argument that a drop in wages raises the rate of capital formation, since the workers' propensity to save wages can be assumed to be smaller than the capitalist's propensity to save profit. We can, therefore, expect a quicker investment in new machines than Ricardo envisaged, so that unemployment will disappear sooner than he believed. The price the workers have to pay for this is a lower level of wages.

According to Malthus, not enough jobs are created either in the short or in the long run to make it possible to re-employ all those who have been displaced by mechanization. He sees no market for the large volume of consumer goods which result from the accumulation of capital. Malthus is thinking here of the economy as a whole, for, as we have seen, he believes partial compensation possible. John Stuart Mill by and large agrees with Ricardo that, in the short run, no significant compensatory employment is likely to come from a drop in prices. In the long run, on the other hand, he is less pessimistic than Malthus, saying that the innovations are made gradually and 'seldom or never by withdrawing circulating capital from actual production, but by the employment of the annual *increase*'.[75] Mill's position is therefore consistent with a long-term version of Say's law.

According to Mill, a static equilibrium is desirable in the long term, and his words written in 1848 are interesting in this connection: 'It is only in the backward countries of the world that increased production is still an important object: in those most advanced, what is economically needed is a better distribution, of which one indispensable means is a stricter restraint on population'.[76] Current discussions about limiting the growth of the economy have more roots in Classical thinking than one might imagine.

3 Marx

Marx had a master then? Yes. Real understanding of his economics begins with recognizing that, as a theorist, he was a pupil of Ricardo.

Schumpeter[1]

3.1 INTRODUCTION

The fact that Ricardo, one of the main pillars of Classical economics, was at the same time an influence on Marx is just one of the paradoxes that confront us when we come to Marx. The main paradox of Marxian economics in the present context is this: while Marx maintains that technical change is accompanied by a decrease in the ratio between human labour (variable capital) and the other, inanimate of 'dead' means of production, such as machinery and raw materials, which he collectively calls 'constant capital', he asserts on the other hand that surplus value is produced only by human labour. Furthermore, he regards work as the source of value, but predicts a large army of unemployed. Again, he speaks of a diminishing rate of profit, while he also forecasts an increasing concentration and centralization of capital. It is not without reason that C. Wright Mills said 'there is no one Marx'.[2]

The paradoxes in Marx's thinking are partly due to the fact that he tried to devise a system that embraced various scientific disciplines. His intention was to depict an integrated model of society in which the interactions between economic and social factors play a central part, and the basic conclusion of which is that capitalism will be destroyed by its internal and inescapable contradictions. Until recently the interest of economists in Marx was limited mainly to his theory of value and his related theory of exploitation. In the last few years, however, more attention has been paid—under the influence of the modern growth theory—to the 'laws of motion' of capitalism, which were more or less explicitly discussed by Marx.[3]

The role Marx attributes to technical change also deserves to be examined here, although taking it out of context risks doing less than justice to Marx's views. A correct assessment of these views is made even more difficult by his tendency to treat definitions of surplus value, rate of profit and the organic composition of capital as if they were behavioural equations.[4] Another source of confusion lies in Marx's frequent reversal of cause and effect, as for example when he says: 'Hitherto we have investigated how surplus-value emanates from capital; we have now to see how capital arises from surplus-value.'[5] A sympathetic interpretation of this is that these two processes are not simultaneous but consecutive. This can be formalized by introducing a time-lag in some of the variables of Marx's theory. It is less satisfactory to view technical change as a function of investment and then view investment as a function of technical change, as was done for example by Higgins[6] in his summary of Marx's theory, while Bronfenbrenner's model[7] of *Das Kapital* did not introduce the time-lags mentioned above, so that he does not do full justice to the dynamic aspects of Marx's argument.

We shall first discuss the role Marx ascribes to technical change and then use Samuelson's model to examine how closely Marx's main conclusions are connected with the type of technical change he assumed in his argument.

3.2 TECHNICAL CHANGE ACCORDING TO MARX

We have seen that Adam Smith viewed mechanization both as the cause and as the effect of the division of labour. We must now return to this concept, since the interactions between mechanization and the division of labour play an important role in Marx's theory as well. Marx looked on Adam Smith as the 'political economist *par excellence* of the period of Manufacture', the hallmark of whose work was the importance he attributed to the division of labour. Marx disagrees with Smith's contention that the invention of machines also results from the division of labour, and says that, while workers were involved in the differentiation of tools in the era of Manufacture (which preceded the setting-up of factories), the invention of machines came from 'learned men, handicraftsmen, and even peasants'.

When discussing the emergence of workshops or 'manufactories', Marx repeatedly stresses the influence which other changes in the structure of production exerted on the division of labour. At the same time, he illustrates—by treating the nature and degree of the division of labour as consequences of dynamic forces—how interwoven are the relationships that characterize the constantly changing circumstances of production. The minimum number of workers the capitalist has to employ in his manufactory is determined by the division of labour, but the advantages of a further division are 'obtainable only by adding to the number of workmen'. When the manufactory reaches a certain size, it turns into the typical form of capitalistic production but at the same time its own narrow technical basis conflicts with the 'requirements of production that were created by manufacture itself'. The manufactory gives rise to the place where mechanical equipment is designed and made: 'This workshop, the product of the division of labour in manufacture, produced in its turn— machines'. This brings to an end a society based on the methods of craftsmen and artisans.

The death of the manufactory marks the birth of large-scale, factory-based industry, the beginning of mechanization and the first step in the advance of capital. Any change, however small, is a revolution for Marx: capital must 'revolutionize the technical and social conditions of the process of work, i.e. the mode of production itself, in order to raise the productivity of labour'. In order to be able to trace the overall evolution back to minor revolutionary changes, Marx is much more concerned with the physical characteristics of mechanical equipment than any of his predecessors. His description encompasses tools and machinery systems (which can be regarded as automated systems), and he concludes that large-scale industry has the necessary technical foundation only when machines are produced with the aid of machines. Unlike the manufactories, the output of the instruments of labour no longer depends on personal human work, and 'thereby the technical foundation on which is based the division of labour in Manufacture is swept away'. A new division of labour, based on the nature of the machines, arises in large-scale industry and the machines dispose of the old division of labour. In the manufactory, the worker uses tools, while in the factory he serves machines. The emphasis shifts from equipping the men to manning the equipment.

The accumulation of capital consists in converting the surplus value into human labour and inanimate means of production, the value of which Marx calls variable and constant capital, respectively. Surplus value comes only from variable capital. The ratio between constant and variable capital is called the organic composition of capital in Marxist terminology. Marx first assumes that the organic composition of capital remains unchanged during the accumulation of capital, so that 'a definite mass of means of production constantly needs the same mass of labour-power to set it in motion'. The demand for labour then increases in proportion to the growth of capital. Since the surplus value produced by the variable capital is added to the capital, the accumulation of capital may lead to a situation in which the demand for labour exceeds the supply on the labour market, and therefore 'wages may rise'. Although this accumulation is beneficial for the workers, their 'relation of dependence and their exploitation' persist. Wages can continue to rise until the increase curbs capital accumulation, but eventually the wage increase exhausts the source of accumulation, and the mechanism of capitalistic production 'removes the very obstacles that it temporarily creates'.[8] Wages drop to the level at which the original rate of accumulation is re-established.

Marx next discusses the case in which the 'vicious assumption'[9] that the organic composition of capital remains unchanged no longer applies. During its evolution, the capitalistic system arrives at a stage where the accumulation is accompanied by regular changes in the organic composition of capital, which are connected with the rise in the productivity of labour: 'The mass of the means of production which he thus transforms increases with the productiveness of his labour'. Owing to this increase in productivity, the constant capital increases and the variable capital decreases, so that the organic composition of capital increases.

As the accumulation progresses, more and more capital is concentrated in the hands of individual capitalists, and the scale of production is enlarged so that the productivity of labour can be increased further. This type of concentration is limited by the growth of rate of 'social wealth' and is characterized by a uniform distribution of the capital between a large number of capitalists who compete with one another. The next, fundamentally different, stage of accumulation is characterized by the 'concentration of capitals already formed, destruction of their individual independence, expropriation of capitalist by capitalist, transformation of many small into few large capitals'. This concentration, called 'centralization' in volume one of the third edition of *Das Kapital*, is not limited by the absolute growth of production and accumulation. Competition eliminates the manufacturers who fail to introduce new methods quickly enough. This competition is carried on through the lowering of prices which is made possible by the higher productivity of labour, and this, in turn, derives partly from the larger scale of production. 'Therefore, the larger capitals beat the smaller.' This marks another revolutionary change in the relationships that characterize production, since centralization ushers in a period in which vast projects such as a railway network can be accomplished. Centralization in this sense is complementary to accumulation, because production can be carried out on a very large scale with the utilization of new inventions and discoveries. 'The absolute reduction in the demand for labour which necessarily follows from this is obviously so much greater the higher the degree in which the capitals undergoing this process of renewal are already massed together by virtue of the centralization movement'.[10]

Nowhere does Marx suggest that the inventors of new machines and methods themselves benefit from this centralization; in fact they often go bankrupt, for it is the 'most worthless and miserable sort of money-capitalists who draw the greatest profit out of all new developments of the universal labour of the human spirit and their social applications through combined labour'.[11] This utilization of new inventions, inherent in centralization, increases the power of capital, and confers at the same time social power.

3.3 DISPLACEMENT OF LABOUR

Marx endorses Ricardo's famous statement that 'machinery and labour are in constant competition' but attaches a more general meaning to it than Ricardo himself did. He also distinguishes between gradual and sudden mechanization, but adds that the workers lose out in the long run in both cases. Marx does not share the optimism of the compensation theory, in any of its various forms. If workers who have been made redundant in one sector do find jobs in another, this is not because the existing variable capital is transformed into machines, but because of new investments. The weak position of the workers arises partly from the need to retrain them for the new job. Only capital formation can create new jobs, and if the organic composition of capital does not change at the same time, the number of people employed rises. According to Marx, this happens when development 'takes a breather'. However, such periods become shorter as the accumulation progresses, for at the same time the organic composition of capital increases, owing to the increasing productivity of labour, and centralization becomes more intense. The reserve army of unemployed is increased, because the redundancy resulting from a replacement of variable capital by constant capital outweighs the compensation resulting from the accumulation of capital, from the greater demand for 'luxury articles' and from the tendency towards the 'unproductive employment of a larger and larger part of the working-class' for example as servants. As the level of wages drops, the aggregate demand for consumer goods drops too, and the incentive to invest decreases. 'Accumulation of wealth at one pole is, therefore, at the same time accumulation of misery, agony of toil, slavery, ignorance, brutality, mental degradation, at the opposite pole, i.e. on the side of the class that produces its own product in the form of capital'.[12]

3.4 MARX AND SAMUELSON

In his analysis of Marx's theory, Samuelson[13] distinguishes between two sectors, the first producing only homogeneous capital goods K, and the second producing only homogeneous consumer goods Y. The production of a unit capital good K in sector 1 takes a_1 units of labour L_1 and b_1 units of capital K_1, while the production of a unit of consumer good Y in sector 2 requires a_2 units of labour L_2 and b_2 units of capital K_2. The coefficients a_1, a_2, b_1 and b_2 express the fixed proportions prevailing in the production and are called technical coefficients. The production in period $t+1$ is assumed to come from the factors of production used up in period t. Taking account of the possibility that labour or capital is abundant in one of the sectors, the production $K(t+1)$, $Y(t+1)$ in the period $t+1$ is given by:

Sector 1: $L_1(t) \geq a_1 K(t+1)$ $K_1(t) \geq b_1 K(t+1)$ (1)

Sector 2: $L_2(t) \geq a_2 Y(t+1)$ $K_2(t) \geq b_2 Y(t+1)$ (2)

The total amount of labour in period t is $L(t) = L_1(t) + L_2(t)$, and the total amount of capital is $K(t) = K_1(t) + K_2(t)$. The production in any given period is limited by the amount of labour and capital available, which means that the total amount of labour in the two sectors is, at most, equal to $L(t)$, and the total amount of capital is, at most, equal to $K(t)$, so that the following relationships hold:

$$a_1 K(t+1) + a_2 Y(t+1) \leq L(t) \tag{3}$$

$$b_1 K(t+1) + b_2 Y(t+1) \leq K(t) \tag{4}$$

Samuelson first assumes that none of the above quantities increases with time, so that each period is an exact replica of the previous one. He also assumes that the amount of labour is fixed, and the amount of capital adjusts itself to it; this reduces Equations (3) and (4) to:

$$a_1 K + a_2 Y = L \tag{5}$$

$$b_1 K + b_2 Y = K \tag{6}$$

These two equations tell us the quantity of capital goods K and consumer goods Y produced ($b_1 < 1$).

Next, Samuelson introduces a price system and denotes the price of the capital good by p_1, the price of the consumer good by p_2, the wage by w and the rate of interest by r. On the assumption that perfect competition prevails and the factors of production are remunerated at the beginning of the period, we now have:

$$p_1 = (wa_1 + p_1 b_1).(1 + r) \tag{7}$$

$$p_2 = (wa_2 + p_1 b_2).(1 + r) \tag{8}$$

From these equations, we can derive the real wage, expressed in terms of consumer goods (w/p_2). Having determined the prices, we can now establish the flow of money in each sector; for sector 1, we have:

$$p_1K = (wL_1 + p_1K_1).(1+r) \tag{9}$$

and for sector 2:

$$p_2Y = (wL_2 + p_1K_2).(1+r) \tag{10}$$

Samuelson then calls $C_1(=p_1K_1)$ and $V_1(=wL_1)$ Marx's constant and variable capital (in sector 1), respectively, and he defines the surplus value S_1 as the difference between p_1K and $C_1 + V_1$. The corresponding relationships for sector 2 are of course $C_2 = p_1K_2$ and $V_2 = wL_2$, and the surplus value is S_2, so that:

$$p_1K = C_1 + V_1 + S_1 \tag{11}$$

$$p_2Y = C_2 + V_2 + S_2 \tag{12}$$

The S values in the two sectors are $S_1 = (C_1 + V_1)r$ and $S_2 = (C_2 + V_2)r$. Samuelson then demonstrates that the degree of exploitation is not the same in the two sectors (i.e. $S_1/V_1 \neq S_2/V_2$), as Marx assumed in volume I of *Das Kapital*. The expressions for these two quotients are as follows:

$$\frac{S_1}{V_1} = \left(\frac{C_1}{V_1} + 1\right)r \text{ and } \frac{S_2}{V_2} = \left(\frac{C_2}{V_2} + 1\right)r$$

Irrespective of the value of r, the degree of exploitation is the same in the two sectors only if the organic composition of capital is the same. Samuelson defines the organic composition of capital as $C:V$. Since, as Blaug[14] said, Marx 'never explicitly defined the so-called organic composition of capital', it is hardly surprising that there are also some other interpretations of this important Marxian concept. Thus Klein defines it as $C/(C+V)$ in his article mentioned earlier (note 4, page 39). In Samuelson's case, the degrees of exploitation are equal if the technical coefficients are such that $a_1b_2 = a_2b_1$. It is easy to show that this condition is also satisfied when the organic composition of capital, as defined by Klein, is the same in both sectors.

The above relationship between the degree of exploitation, the organic composition of capital and the rate of interest can easily be written to obtain the well-known Marxian tautology that the rate of interest r is equal to the rate of profit $S/(C + V)$. Marx used this identity to show that an increase in the organic composition of capital, caused by the accumulation of capital, must lead to a decreasing rate of profit at a constant degree of exploitation.[15] Like many before him, Samuelson points out that this argument is untenable and maintains that the decrease in the rate of profit cannot be derived in this way, for one must also specify the effect of the capital accumulation on the technical coefficients a and b, as well as on the demand for, and the supply of, labour and capital.

Samuelson begins by discussing the famous Marxian contradiction that as the rate of profit decreases the real wage decreases or remains the same. It can easily be shown that, with given technical coefficients a and b, the real wage w/p_2 in fact increases when the rate of profit (or interest) r decreases under the influence of accumulation.

This result is obtained when a given level of technology is assumed, and Samuelson questions the relevance of the proposition because of the technical change with which Marx operates. Samuelson believes that a technical change must be an improvement, since otherwise it would not be introduced in an economic system in which perfect competition prevails. 'Marx cannot repeal the valid part of Adam Smith's law of the Invisible Hand, for its validity depends on the existence of numerous avaricious competitors'.[16] According to Samuelson, a new set of technical coefficients a' and b', with r' ($< r$), leads to an increase in the real wages w/p_2.

To throw some light on the background of Samuelson's argument, it is necessary to examine the connection between the real wage w/p_2 and the rate of interest r in this model with the aid of a wage-interest curve, shown in Figure 3, which assumes a given state of technology (a, b) and has a negative gradient, i.e. it falls from left to right. The corresponding wage-interest curve for the new (smaller) technical coefficients a', b', characteristic of the new technology, will lie to the right of the original curve, and is shown in Figure 3 (curve 2).

If we start from point A, a lowering of the rate of interest r

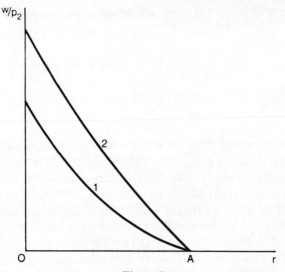

Figure 3

will—assuming perfect competition and sufficiently quick responses—lead to the introduction of new technology, which implies a higher real wage. In the simple situation depicted here, the new technology will be put into use at any rate of interest (apart from point A). The complications which arise when the wage-interest curves intersect at a point other than A will be discussed later (Sections 9.2 and 9.3). The wage-interest graph is a straight line if the organic composition of capital is the same in the two sectors and there is no discounting of the wage sum.

To examine the effect of the accumulation of capital, K and L are assumed to increase at a growth rate m. Samuelson also assumes that all the available means of production are employed. By writing the relationship $K(t+1) = (1+m)K(t)$ out in full for the two sectors, we obtain:

$$a_1(1+m). \ K(t)+a_2 Y(t+1) = L(t) \qquad (13)$$

$$b_1(1+m). \ K(t)+b_2 Y(t+1) = K(t) \qquad (14)$$

It is possible to derive from these equations the growth paths for both sectors at a given labour input $L(0)$ for the initial situation. If the organic composition of capital is the same in both sectors, i.e. if $a_1 b_2 = a_2 b_1$, the corresponding growth paths are given by:

$$Y(t+1) = \frac{1 - b_1(1+m)}{a_2} \cdot (1+m)^t . L(0) \qquad (15)$$

$$K(t) = \frac{b_2}{a_2} (1 \times m)^t . L(0) \qquad (16)$$

Samuelson then tackles the situation where capital and labour do not grow at the same rate, capital accumulation increasing faster than the supply of labour. We can examine this situation by assuming that the supply of labour remains equal to $L(0)$, so that the rate of growth of labour is zero. If capital accumulation then takes place, there will soon be an excess of new capital goods in this model because fixed proportions are assumed. As soon as the ratio between capital and labour exceeds b_2/a_2 in a given period, the price of the capital goods, p_1, drops to zero. Only capital goods needed to utilize the fixed supply of labour will then be produced in the future. The first sector, where capital goods are made, shrinks, so that $K(t+1)$ no longer exceeds $(b_2/a_2).L(0)$. Samuelson's conclusion is that accumulation is impossible in this Marxist model: 'there can be no technical "deepening of capital" in it'. Samuelson believes that the rate of profit r can still easily fall, since individuals will carry on saving—even though no new investments can be made. The lower the rate of profit r, the higher the real wage. Samuelson says that every capitalist 'in trying to save and increase his own profits ends up killing off the total of profits in favor of the workers'.[17]

If there are different technical possibilities (a, b), a drop in r can lead to the introduction of new technology. If the new technology is of the labour-saving type, the real wage will rise to a smaller extent than in the case when no alternative technology is available. However, since the new technology is introduced because r is lower due to capital accumulation, the real wage will be higher than under the original technology. If a constant rate of growth m is assumed, a relationship can be found between per-capita consumption, c, and this rate of growth, which parallels the relationship between the real wage and the rate of interest r, mentioned earlier.[18] These parallel relationships indicate that the per-capita consumption is equal to the real wage, provided that the rate of growth m is equal to the rate of interest (see Section 12.4).

The elegant and interesting feature of Samuelson's model is that it manages to incorporate essential features of Marxist economics in a neo-classical framework in which all the dissonant elements combine into a harmonious whole. In Samuelson's picture, nothing remains of the conflict of interests between the capitalists and the workers, and no conflicts arise from the one-sided accumulation of wealth and the concentration of capital. It is a harmonious world in which perfect competition ensures that no-one acquires a position of dominance in the market. Such a state of affairs is found also in Morishima's model,[19] which can be regarded as a generalized version of Samuelson's.

Both models lead to the conclusion that technical change does not produce any significant disturbance. The decrease in the rate of profit, caused by the accumulation of capital, makes entrepreneurs use alternative, labour-saving techniques; the price mechanism ensures that labour and capital are employed in accordance with the technical coefficients of the production processes, and in any case the real wage does not fall. Marx has thus been taken over and emasculated by Classical economics—an economics, however, that uses modern mathematical apparatus.

This picture diverges sharply from Schumpeter's account of Marx. According to Schumpeter a capitalistic system does not expand 'in a steady manner' and instead there is a permanent endogenous upheaval caused both by new products and new methods of production, the whole system being characterized by constant change. New changes arrive before the adjustments to previous phenomena are complete. Competition does not operate in a stationary equilibrium; new products and methods compete with the old ones not on an equal footing but with a definite advantage which ensures that the old ones are ousted. Companies must therefore embrace the new—i.e. they must invest—otherwise they go under. Those companies that survive accumulate capital and finance this accumulation from profits. Schumpeter concludes that Marx had a better insight into industrial change than 'any other economist of his time'.[20]

3.5 ACCUMULATION, PRODUCTIVITY OF LABOUR AND THE ORGANIC COMPOSITION OF CAPITAL

The relationships between these three factors in Marxian theory will now be examined with the aid of the stationary variant of Samuelson's model.[21] This is appropriate because it retains Marx's differentiation between two sectors, starts with fixed proportions, and gives a clear description of the Marxian concepts (without, however, being a good representation of Marx's theory as a whole).

The table below gives the expressions for the productivity of labour, the capital coefficient and the organic composition of capital for both sectors, the capital coefficient being the ratio between the capital and the volume of production. Our analysis is based on a given rate of interest r. Since the wage-interest curves are shifted to the right by technical change in these examples, the new technology yields a higher wage than the old one at any given rate of interest. If these curves intersect, the assumption of a given value for r is inadmissible, since in that case the choice of technique is dependent on the value of r, and it can no longer be assumed that the new technology is invariably preferred to the old one.

	Sector 1	Sector 2	Total
Productivity of labour	$1/a_1$	$1/a_2$	$\dfrac{1-b_1+b_2}{a_2(1-b_1)+a_1b_2}$
Capital coefficient	b_1	b_2	$\dfrac{b_2}{1-b_1+b_2}$
Organic composition of capital	$\dfrac{b_1(1+r)}{1-b_1(1+r)}$	$\dfrac{a_1b_2(1+r)}{a_2[1-b_1(1+r)]}$	$\dfrac{a_1b_2(1+r)}{[1-b_1(1+r)][a_2(1-b_1)+a_1b_2]}$

Marx first assumes that the organic composition of capital remains constant, and we have added to this the assumption that it is the same in both sectors (which means that $a_1b_2 = a_2b_1$). The total organic composition of capital is then equal to that in the two sectors, i.e.:

$$\frac{b_1(1+r)}{1-b_1(1+r)}$$

If this expression remains constant, then the input coefficient b_1 must also be constant and one can also assume that the capital coefficient in sector 1 remains the same. It is reasonable to assume that Marx meant that the capital coefficient in sector 2 (b_2) was also constant. In this situation, a constant organic composition of capital is also obtained if one starts with a constant capital coefficient. Since $a_2 b_1 = a_1 b_2$, one still has a choice as far as the coefficients a_1 and a_2 are concerned.

Marx presumably assumed that both remained constant in the case discussed by him; however, this is by no means necessary, and the productivity of labour might rise equally in sectors 1 and 2, so that $a_1(t) = a_1(0).e^{-kt}$ and $a_2(t) = a_2(0).e^{-kt}$ (where $k > 0$). We are dealing here with technical change brought about, for example, by new inventions while the partial and the total organic composition of capital remain unchanged. With a constant rate of growth—in Marxist terminology, a constant degree of accumulation—the situation may lead to the dismissal of workers, depending on the magnitude of k. Marx apparently did not consider this and assumed just the opposite, i.e. that at a constant organic composition of capital the accumulation leads to an increase in the demand for labour. (It must be assumed that the set of input coefficients remains unchanged.)

If the two partial organic compositions of capital are now assumed to be different, then the expression for their sum involves all four input coefficients a_1, a_2, b_1 and b_2, which are again assumed to be constant for the sake of simplicity. The productivities of labour and the capital coefficients are then also constant, and so no technical change takes place. However, one can visualize types of technical change that leave the total organic composition of capital unchanged. For example, the productivity of labour can again be assumed to grow equally in the two sectors at a rate of growth k. In the case of technical change of this type, three of the capital coefficients remain constant. The displacement of labour then depends on the degree of accumulation and on the growth rate k. However, there is a second interesting possibility, in which the coefficients a_2 and b_2 decrease to the same extent, for example on the assumption that $a_2(t) = a_2(0).e^{-kt}$ and $b_2(t) = b_2(0).e^{-kt}$.

The partial and total organic compositions of capital remain constant in this case too. Another type of technical change affects the second sector only. In this case, the productivity of labour increases in sector 2, while remaining constant in sector 1. Furthermore, the capital coefficient stays constant in the first sector, while decreasing in the second. The total capital coefficient also decreases. Displacement of labour can occur in sector 2, depending on the degree of accumulation and on the extent to which the productivity of labour rises in this sector.

All this suggests that technical change can be distinguished from accumulation. And we have seen that the sectoral analysis of an economic system developed by Marx leads to the recognition of various types of technical change.

That an incomplete distinction between the organic composition of capital and the capital coefficient can lead to confusion appears from an analysis of the typical Marxian hypothesis in which the organic composition of capital increases.[22] Marx and many of his followers mention the increase in the organic composition in the same breath as the increase in the productivity of labour. We shall now assume an increasing productivity of labour and examine how far an increasing organic composition of capital follows from it. The first difficulty concerns the choice of the factors, or input coefficients, responsible for the rise in the productivity of labour. The expression for the total productivity of labour again contains all four input coefficients, so that in principle each of the four coefficients is a possible choice. Yet it is practically compulsory in Marxian economics to operate only with the coefficients that determine the labour input, i.e. with a_1 and a_2, which determine at the same time the productivity of labour in sectors 1 and 2. It is assumed that the increase in the productivity of labour is the same in the two sectors, i.e. that $a_1(t) = a_1(0).e^{-kt}$ and $a_2(t) = a_2(0).e^{-kt}$. The coefficients b_1 and b_2 remain constant. We again assume that the organic composition of capital is the same in both sectors (i.e. $a_1b_2 = a_2b_1$), so that it is given by:

$$\frac{b_1(1+r)}{1-b_1(1+r)}$$

It can be seen immediately that the organic composition of capital is not affected by an increase in the productivity of labour in the two sectors, which is of course accompanied by an increase in the total productivity of labour. Although Marx therefore could not have been thinking of this state of affairs, that does not detract from its feasibility.

It is natural to assume now that the organic composition of capital is different in the two sectors; if the different decreases in the coefficients a_1 and a_2 are taken into account, the total organic composition of capital is given by:

$$\frac{a_1(0)b_2(1+r)}{[1-b_1(1+r)][a_2(0)(1-b_1).e^{(k-l)t}+a_1(0)b_2]}$$

It is assumed here that the increase in the productivity of labour is different in the two sectors, so that $a_1(t) = a_1(0).e^{-kt}$ and $a_2(t) = a_2(0).e^{-lt}$, with $k > 0$ and $l > 0$.

The organic composition of capital will thus increase if $k < l$. When the productivity of labour increases in sector 1 more slowly than in sector 2, the technical change is accompanied by an increase in the ratio between the constant and the variable capital, i.e. in the organic composition of capital. It is interesting that the partial organic composition of capital remains constant in sector 1, while it increases in sector 2. This situation, which was implied by Marx,[23] does not affect the capital coefficients.

In the cases discussed above, technical change always leads to such a shift in the wage-interest curve that no switch points occur. This means that at each given rate of interest r, the latest technology is used in order to maximize profits in a situation of perfect competition. Since only circulating capital goods exist in this model, the new technology can be introduced in a smooth and flexible manner, although the choice of the best technique is important from the point of view of social welfare insofar as the new technology requires less labour.

3.6 TECHNICAL CHANGE AND THE LABOUR MARKET

The following can be said about the displacement of labour. On the one hand, the constant rate of accumulation results in a regular demand for labour, while on the other hand redundancy occurs first in sector 2, depending on the magnitude of k and l. As is well known, Marx postulates that in the long run the redundancy due to the increase in the productivity of labour exceeds the demand for labour caused by the accumulation of capital. Steindl rightly remarks that this is based on the assumption of a given rate of accumulation.[24]

We shall now look at the effect of technical change on the labour market using Samuelson's model once more. It has already been shown with the aid of this model that the productivity of labour may increase even when the organic composition of capital remains constant. This warns us against attributing redundancy exclusively to a change in the organic composition of capital, as was done for example by Gottheil.[25] Even though Marx implied no qualitative change in the case of a constant organic composition of capital, the analysis of the situation in which the organic composition of capital changes will be useful only if the type of technical change that causes this is taken into account.

We first assume that the productivity of labour in sectors 1 and 2 changes thus: $a_1(t) = a_1(0).e^{-kt}$ and $a_2(t) = a_2(0).e^{-lt}$, where $k < l$, so that the organic composition of capital increases. We also assume that the working population increases in the two sectors thus: $L_1(t) = e^{nt}L_1(0)$ and $L_2(t) = e^{nt}L_2(0)$, and that capital is accumulated thus: $K_1(t) = e^{mt}K_1(0)$ and $K_2(t) = e^{mt}K_2(0)$, where m is the rate of accumulation. We further assume that fixed proportions prevail and that in the initial state there is full employment, so that $L_1(0) = a_1(0).K(0)$ and $L_2(0) = a_2(0).Y(0)$. The degree of employment, i.e. the ratio between the demand for labour and the supply of labour in sectors 1 and 2 is then given by μ_1 and μ_2, respectively:

$$\mu_1 = \frac{\dfrac{a_1(0)}{b_1}K_1(0).e^{(m-k)}}{\dfrac{a_1(0)}{b_1}K_1(0).e^{nt}} = e^{(m-k-n)t}$$

$$\mu_2 = \frac{\dfrac{a_2(0)}{b_2}K_2(0).e^{(m-l)t}}{\dfrac{a_2(0)}{b_2}K_2(0).e^{nt}} = e^{(m-l-n)t}$$

A degree of employment of 1 implies full employment. For this to occur, the degree of accumulation m must be equal to $k+n$ in sector 1, and to $l+n$ in sector 2. Since $k < l$, redundancy occurs in any case in sector 2 if sector 1 is fully manned, the degree of accumulation being too low for there to be jobs for all the people in sector 2. Unemployment in both sectors is, of course, entirely feasible; this happens when $m < (k+n)$.

It is now quite clear that the displacement of labour in Marx's theory is based on two postulates, namely a given rate of accumulation and a given type of technical change which does not influence the input coefficients b_1 and b_2 and thus the capital coefficients. If the rate of accumulation m increases in the course of time, e.g. as a result of technical change, then redundancy will be less. Such an increase in the rate of accumulation is possible, however, only if as well as the capitalists the workers also contribute to capital accumulation by saving part of their income. But the conventional Marxist view is that only capitalists re-invest part of the surplus value (while spending the rest on consumer goods), so that an increase in their propensity to save and invest implies a drop in the propensity to consume. In this respect, Marx's remarks about compensation through a higher demand for luxury goods are not entirely consistent with his central theory.

In the case of a different type of technical change, for example one which leads to a change in the input coefficients b_1 and b_2, the degree of accumulation m can also increase when more capital becomes available through technical change of this type. In this case, too, redundancies may occur on a smaller scale than Marx assumed, since the employment of workers need not be hampered by a lack of capital.

Comparison between Classical thinking and Marxism on the displacement of labour shows that whereas, in the former, technical change does not lead to a permanent under-utilization of the available labour force, in the latter it leads to a certain disproportionality, with the capital goods being always fully utilized, but the labour not. The Classical economists generally assume an adequate rise in the aggregate demand—either because of a fall in prices or an increase in disposable income—which proceeds proportionately, since no-one holds a position of power in the system. In Marx's example this adequate increase in demand is lacking, since technical change leads to a concentration of power on the supply side which prevents the commensurate increase of demand. For this reason, Marx's technical change is not a factor that compensates for the diminishing returns of the Classical economists, but something that when it is fully effective, has a negative influence on the real wages of the workers. It should be noted, furthermore, that Marx's analysis of redundancy and partial compensation is a step forward from Say's incomplete analysis, already discussed, in that his sectoral approach brings the whole system into play, which is a move towards a general equilibrium analysis. The nature and extent of redundancy vary with the type of production function used in the argument, as well as with the changes brought about in this function by technical change in the course of time. The important point about Marx's theory in this connection is that it operates with fixed proportions and inventions which are invariably labour-saving.

3.7 TYPES OF TECHNICAL CHANGE

We can conclude from the argument presented so far that this aspect of Marxian economics is quite consistent if a certain type of technical change is assumed. It is therefore important to differentiate between the changes in the productivity of labour in the different sectors, as expressed by the labour coefficients a_1 and a_2. It is not difficult to see why Blaug thought the weakness of Marx's growth theory lay in the disappearance of 'investment prospects not because there have been too few labour-saving improvements, but because there have been too many'.[26]

Modern theories of technical change can help clarify what type of technical change Marx was thinking of. Without going into details of the classification at this stage, we shall examine one type of technical change—that labelled 'neutral' by Hicks and Harrod.

In his book on the theory of wages, Hicks[27] follows Pigou[28] on the nature of inventions and their influence on the relationship between labour and capital. Pigou and Hicks distinguish between capital-saving, labour-saving and neutral inventions, according to the effects these inventions have on the marginal product of labour and capital; if they raise the marginal product of labour more than they raise the marginal product of capital, they are labour-saving, and if they raise the marginal product of capital more than they raise the marginal product of labour, they are capital-saving. 'Neutral' technical change takes place, according to Hicks, when the marginal products of labour and capital are affected to the same extent. Though a precise definition calls for a production function, we can show that one type of technical change, which has been discussed in connection with Marx, can be regarded as 'neutral' in Hicks' sense.

The fact that, as shown in Section 3.5, a constant organic composition of capital is possible when the labour coefficient and the capital coefficient decrease to the same extent in sector 2 (with a different organic composition of capital in the two sectors) means that technical change has an equal effect on the coefficients a_2 and b_2 in the sector producing consumer goods, and thus the ratio between the marginal product of labour and that of capital remains unaffected, so that sector 2 does indeed experience a neutral technical change (in Hicks' sense).

Harrod uses a different approach and defines neutral technical change as 'one which, at a constant rate of interest, does not disturb the value of the capital coefficient'.[29] This is precisely what happens in the second important case already discussed, in which the coefficients a_1 and a_2 experience an equal or different reduction under the influence of technical change. In the characteristic Marxian case, the organic composition of capital increases because the productivity of labour does not rise as fast in sector 1 as it does in sector 2. The capital coefficients b_1 and b_2 and the rate of interest r remain constant, so this is neutral technical change in Harrod's sense.

The conclusion that Marx was mainly thinking of neutral technical change of Harrod's type gains support from his theory of employment. Harrod's neutral technical change is also called labour-augmenting, since its effect is equivalent to an increase in the working population. If fixed proportions are assumed in Marx's theory, it is immaterial whether the effect on capital (which is the critical factor) results from a growth in the working population or from a neutral technical change of Harrod's type, since both boil down to the same thing—an increase in the labour factor of production. In Marx's theory, unemployment results from the fact that redundancy caused by neutral technical change of Harrod's type outstrips the demand for labour caused by the accumulation of capital.

3.8 A FURTHER INTERPRETATION OF MARX

Our analysis has so far been based on a given rate of accumulation and the assumption that the new technique takes over entirely from the old method of production. To study the effect of technical change on the partial and total composition of capital, it has also been assumed for analytical purposes that the rate of interest (or profit) is also given.[30]

One should notice here that, in this analytical framework, the application of a new technique implies a rise in the real wage rate.[31] Whether the real wage in fact rises depends to a great extent on the labour market, and one should not forget that the whole apparatus based on wage-interest curves forms an open system that can be closed via different routes. Since the supply of labour depends on population growth, the assumption of the latter is essential. The demand for labour depends on the actual rate of accumulation of capital, and if this rate is relatively low its negative effect on the real wage will more than counterbalance the positive effect of a new technique. The change in real wages also depends on prices. If during the later stages of accumulation producers maintain the level of prices despite the application of new techniques, the real wages will not rise. Finally, there is the question of the relationship between the employed and the unemployed workers and its effect on real wages.

According to Marx substantial unemployment follows as an inevitable phenomenon from the given rate of accumulation and from the application of new techniques. Depending on the magnitude of the effects exerted by these factors on the real wage rate, the latter may, on balance, rise, though it need not necessarily do so.

A supplementary and further interpretation of Marx is that the real wage rate is kept constant, because it is equal to the absolute minimum necessary for survival.[32] The application of a new technique is then accompanied by a rise in the rate of profit, and if the rate of accumulation is raised by an increase in the rate of savings and in the level of investment, then an increase in unemployment is no longer certain. It is also questionable whether the assumption of a constant real wage rate can be maintained and if this rate increases, accumulation is restricted. This argument may form the starting point for the postulation of an endogenous cycle of the rate of accumulation, as demonstrated by Eagly.[33] During the boom periods of such a cycle, the typical Marxian forecasts of an army of unemployed seem to disappear, but in the long run they reappear undiminished. As Mattick remarks, technical development in Marx's theory always displaces labour, but the level of employment depends on the rate of capital formation.[34]

3.9 CONCLUSIONS

Samuelson's model is useful for an analysis of Marx, but it does not provide a full description of Marx's theory. The only manifestation of technical change in the model is a formal change in the technical coefficients. According to Marx, a rise in the productivity of labour (which, in his view, raises the organic composition of capital) can result not only from the introduction of new machines, but also from a further division of labour, an improvement in organization or an increase in the scale of production.

Samuelson's model ignores also the Marxian connection between accumulation, concentration, centralization and the increase in the productivity of labour, and it says nothing about the class struggle. However, we have been able to discuss the displacement of labour by developing Samuelson's model further, by considering redundancy as the product of two forces with opposite effects on the level of employment. The relative magnitude of these forces determines whether there is an increase in unemployment.

Samuelson's model also gives an incomplete picture of Marx's ideas because it explicitly assumes perfect competition, and while there are references in Marx's work to justify this assumption it is very doubtful whether Marx was thinking of a situation where producers have no control over prices, because this would be incompatible with his picture of the future of capitalism, dominated by the increasing accumulation of power by capitalists. Rather, Marx's belief that competition encourages the application of new inventions—and that in the ensuing turmoil some prosper and others flounder, with the eventual disappearance of smaller companies—points towards oligopoly and monopolistic competition. Marx was not entirely consistent in so far as he introduced into his description of the changes in the capitalistic mode of production certain oligopolistic elements at a stage of capitalism which, even according to him, was still mostly characterized by perfect competition.

Baran and Sweezy believe that Marx definitely had in mind a type of capitalism in which monopolistic forms of competition played an important role but he never attempted to analyse the situation in detail, mainly because he expected that capitalism would come to an end long before 'the unfolding of all its potentialities, well within the system's competitive phase'.[35] Baran and Sweezy's view illustrates the difficulties involved in summarizing Marxian economics in single model. What is important in this connection is to decide to what extent Marx thought of technical change, accumulation and concentration as phenomena evolving from within the capitalistic system, or as phenomena determined by events taking place outside the system. The more endogenous the change is, the more arbitrary is the demarcation of the successive stages of capitalism.

To begin with, mechanization and the accumulation of capital are, for Marx, exogenous phenomena. While the degree of accumulation of capital is not explained in detail by Marx and is thus clearly an exogenous effect, mechanization—once it has begun—is partly explained by competitive factors. And the progressive division of labour, increases in scale and organizational changes to a large extent follow automatically from mechanization. These forces interact and, together with further accumulation of capital, produce concentration and centralization. The inner conflicts which characterize capitalism become more and more pronounced and are closely connected with the growing catalytic effect technical change has on capitalism.

While the Classical economists saw the introduction of new machines as an external disturbance to which the system would eventually adapt itself and whose influence depended mainly on how gradually technical change took place, Marx emphasized the effect of mechanization on production and the factors of production in a wider sense, and so paid more attention to the endogenous elements of technical change. Marx was the first economist to realize fully the significance of technical change for economics and society. In particular, he realized the significance of the invention and application of new machines for the division of labour, large-scale production, the creation of new products, and the phenomena of concentration and centralization. He developed a full, coherent picture of the industrial revolution and described its consequences for both the individual and mankind.

4 Technical development up to 1900

*Modern Industry never looks upon and treats the
existing form of a process as final. The technical basis
of that industry is therefore revolutionary, while all
earlier modes of production were essentially
conservative*

Marx

4.1 SLOW BEGINNINGS

This account of technical development must necessarily be brief
and in some respects incomplete. We shall concentrate on the
recurrent features of the process, though noting that they never
recur in exactly the same form. An attempt will be made to
establish to what extent economic theory responds to the actual
pattern of technical change in society. We shall restrict the
analysis to western civilization.[1]

Man has worked with his hands from time immemorial;
slowly he learned to shape stones into tools, which were better
for hunting, fishing, making clothes and building huts. The step
from unhewn stones to worked ones was an enormous one, and
the stone-axe and knife appeared only after a long time. The
discovery of fire led to ways of keeping warm and cooking food;
meat could be kept longer and rudimentary stocks built up.
The experience gained in cooking and roasting were later to be
utilized in metallurgical processes and firing pottery. The less
time was needed for the acquisition of food, the more time could
be devoted to improving tools and weapons. It thus became
possible to dig for example for salt. Though there was hardly
any division of labour yet, goods were bartered—together with
precious stones—and quite considerable trade routes were
established. Certain animals were domesticated and slowly
agriculture evolved. More durable dwelling places were built,
and more or less permanent settlements were founded. The
invention of the wheel was enormously significant, being
overshadowed in importance only by the invention of writing.

Important civilizations sprang up along the Nile, Tigris and Euphrates around 3000 BC. Since the Egyptian and the Mesopotamian area had a different geography they experienced different forms of technical development and political organization. But since the rivers in these areas behaved in the same way, both the Egyptians and the Mesopotamians developed irrigation techniques to put the excess water into use for the production of essential commodities. According to Forbes,[2] the technological character of these irrigation devices—dikes and canals—also influenced the political and social evolution in these countries. The scale of the activities required cooperation, which was enforced by the authorities. The autocratic form of government also played an important part in the building of the pyramids, motivated by religious reasons. Simple ways of working iron, copper, silver and gold already existed at this time. However, with the collapse of the political superstructure, the development of the technology supporting these civilizations also petered out.

In Graeco-Roman times—between 900 BC and 400 AD—there arose, besides the basically agricultural communities, a few towns, which engaged in a brisk trade. One of the most interesting Greek inventors was Ktesibios, who lived in Alexandria around 270 BC and who constructed a clepsydra, or water clock. Water flowed into a tank, with an overflow at the side and a hole at the bottom, through which the water escaped at a constant rate. The water flowed into a tube fitted with a float that had a pointer, which moved against a scale showing the hours.

While Ktesibios' name has been rescued from anonymity, we do not know the names of many other inventors like him. Drachmann[3] has discussed the gradual modifications which evolved in the technique of threshing, which led to the introduction of water mills based on the water-wheel. Although labour was not scarce, partly because of the use of slaves, the aim of many inventions was still to save work, and the water-wheel showed that human labour could be replaced by energy from other sources. To satisfy the needs of armies new weapons were invented. The Romans also constructed an extensive network of roads, which enabled them not only to rule over a vast area but to develop trade. Many of these roads remain and are a tribute to the Romans' organizing powers.

In the Middle Ages technical development proceeded at a slow but discernible pace and, most important, people became aware of the possibility of a steady advance; they foresaw the aeroplane, the motor-boat and the car. The road lay open to the Renaissance, because there were people who could see what was as yet invisible, believed in what they saw, and put their beliefs into practice.

The technology of the Renaissance is illustrated by architecture, printing, gunpowder, and Italian glass, earthenware and metal articles. Techniques spread quickly, but were rarely the same in two places. Besides water, wind was also harnessed, and windmills became common in places like Holland. Iron and steel were widely used for making tools, knives, weapons, chains, clocks, and agricultural implements. In the textile field the spinning wheel was improved. The invention of gunpowder had wide repercussions. Waging a war became so expensive that only monarchs could afford it; the importance of feudal lords declined, while national states became steadily more important.

Equally important was printing, the consequences of which are difficult to overestimate. The invention of printing is usually attributed to Johann Gutenberg around 1440. Others attribute it to the Dutchman Laurens Coster which suggests that some inventions were made by two or more people independently. Hall[4] points out the synthesis of various previously known elements in the invention of printing—paper, ink and movable type. Paper was available; the principle of the press—already known—merely had to be applied to pressing and printing the ink on paper; the ink had to be mixed with linseed oil; and finally, somehow the letter types had to be kept together to achieve a reasonable speed of printing.

One might have expected that printing would have brought together science and technology, which had until then been developing along separate lines. But this did not happen. Technology has always been the domain of the practical man, the *Homo faber*, who passed on—often by word of mouth—his knowledge of stone, wood, iron, steel, water, wind, wool, paper and ink, his aim being to create new objects—like the wheel—

without knowing more abstract properties such as their chemical constitution. A practical man does an experiment to confirm what he already knows; a theoretical man does it to find out something he does not know. Up to about 1800, technical development was in the hands of practical men. Admittedly printing led to the recording and disseminating of numerous technical procedures, but the authors of these books left the creation of the new to the creators of the old—the craftsmen.

As time went on, cooperation between science and technology slowly improved. Gradually technical development came to be regarded as the result not so much of accidental discovery but rather of patient, systematic work. This realization raised the status of technology. Although scientists now became more interested in technology and there was more contact between science and technology, the impact on actual technical processes was still small. Changes in agriculture, for example, proceeded at their own slow pace.

The development of metallurgy in the seventeenth and eighteenth centuries illustrates how technical knowledge helped chemical analysis. Ores of the oxide type were reduced to metals by heating them first with wood and later with coal, which oxidized the latter. This process was first used for making metals, and only much later was its scientific basis understood. Similarly, ores could be analysed for silver and gold with the aid of chemical reactions (such as that between silver and nitric acid), without the analyst knowing the chemical basis underlying the method. Analytical methods were devised to detect precious metals in various ores. The three iron products—cast iron, wrought iron and steel— were produced before it was known that it was their carbon content that distinguished them. Considerable advances were made in casting and alloying before chemistry could give a sound explanation of the phenomena involved.

Science and technology were combined by the instrument-makers such as Galileo (1564–1642), who realized the importance of the telescope, developed in Holland, for studying the sky. Not only telescopes but also microscopes and air pumps were developed in his lifetime. The precision, efficiency and the weight of the apparatus increasingly became the important features. The road from theory to application grew longer, and requirements became both more demanding and more specialized. Science and technology became increasingly interdependent.

Not only were science and technology coming together: in Britain inventions increasingly found technical and commercial application. The inventions made in the textile industry during the Industrial Revolution (1750–1830) were particularly spectacular. The first mechanical spinning-machine was invented by John Wyatt or Lewis Paul, the spinning jenny by Hargreaves, the flying-shuttle by Kay, and the power-loom by Cartwright. The great enthusiasm with which inventors went to work on improving textile-making processes inevitably led to simultaneous inventions and cases of patent infringement, such as the well-known ones involving Arkwright.[5]

4.2 CHARLES BABBAGE

We shall now examine an account which the contemporary inventor, Charles Babbage (1792–1871) gave of the rapid development[6] taking place not only in the textile industry but also in other fields in the nineteenth century. Babbage was a professor of mathematics at Cambridge and invented the computer, though this invention did not contribute to technical development in the last century. He also studied economics and was one of those rare people who combine technical knowledge with social involvement.

Babbage professed that the aim of his book *On the Economy of Machinery and Manufactures*[7] was to describe the advantages of tools and machines, the first being powered by hand, the second by steam. He further distinguished between machines used merely as a source of power and machines which also 'execute work'.[8] According to Babbage, the rate of work in non-mechanized processes is slow, because—other factors being equal—the human body gets tired more quickly when it has to perform the same act over and over again. The rate of production can be effectively increased only by mechanization. Babbage mentions the steam-engine, without which many of the mines could not be worked profitably. The telegraph is also a machine 'for conveying information over extensive lines with great rapidity'. A system of signals was invented in Liverpool about 1830 by which each merchant could 'communicate with his own vessel long before she arrives in the port'. One of the greatest advantages of mechanization lies in what Babbage calls the extension of the time during which a force stays in operation. He mentions the watch as an example of something which can be wound up in a few seconds and carries on working all day, and he gives a number of examples illustrating how the speed of natural processes can be increased by inventions. He discusses in detail the methods developed for monitoring and regulating manufacturing processes. He attaches particular importance to automatic signals indicating a fault or breakdown in these processes, and in this respect anticipates the theory of computerized technology. He then goes on to discuss the saving in raw materials and the greater precision and uniformity made possible by mechanization. This leads to a discussion of all kinds of techniques, such as pressing, casting and stamping, by which identical objects can be made in large numbers—lead pipes and vermicelli for example.

After dealing with a questionnaire for empirical investigation in the field of mechanization, he discusses the economic aspects of the latter. He attributes great importance to the statistical investigation of how many new customers an entrepreneur will acquire 'by a given reduction in the price of the article'. He also makes an important distinction between 'making' and 'fabricating', the difference being one of scale: 'making' is the production of a single item, while 'fabrication' is the production of many identical ones. The important point is that if a 'maker' wants to become a 'fabricator' or manufacturer, 'he must attend to other principles besides those mechanical ones on which the successful execution of his work depends'.[9] This involves in particular the continuous and indefatigable pursuit of measures that raise the efficiency of his production, in order to keep up with his competitors. Sooner or later a new production method will be adopted by other manufacturers, so that he will lose out to his competitors unless he keeps alive his innovating spirit and effects cost reductions. In this case, technical development is an important cause of the drop in prices, although partly masked by inflation.

Babbage agrees on the whole with Adam Smith about the

division of labour, and in particular he too regards the invention of tools and machines as something which depends on the division of labour. Like Gioja,[10] Babbage stresses that the advantage of the division of labour is that it permits the manufacturing process to be adapted to specific characteristics of the individual workers. Babbage's main contribution is that he put forward factual proposals for applying the division of labour in the field of mental work. He describes in detail how a machine can be used to find the squares of integers, so as to avoid the 'loss arising from the employment of an accomplished mathematician in performing the lowest processes of arithmetic'.

Babbage believes that the ability to invent is not so rare in itself, but many inventors have 'failed from the imperfect nature of the first trials', and 'a still larger portion failed only because the economy of their operation was not sufficiently attended to'. He adds however that truly great inventions are indeed rare: 'Those which command our admiration equally by the perfection of their effects and the simplicity of their means are found only amongst the happiest productions of genius'. Inventions and improvements to machines and devices should first be presented in the form of drawings or blueprints, although this of course cannot be done where the central issue lies in chemical and physical characteristics that cannot be illustrated. In these cases some experiments should be carried out, for only then can new machines be constructed and tested. Failure at this stage can be attributed to shortcomings at the previous stages of the inventive process. Failure is often due to paying insufficient attention to the fact that 'metals are not perfectly rigid but elastic'. Related to this is the fact that many ideas can be put into practice only when the state of technology is advanced enough to cope with them. Babbage mentions printing, and stresses that the available technical knowledge should be reviewed from time to time when 'the art of making machinery has received any great improvement', in order to see whether it is possible to put into practice previous, theoretically feasible ideas.

However 'the invention may still fail'—even though a blueprint has been prepared, all the parts of the machine function perfectly, and the product comes up to expectations—

because of unfavourable cost and profit figures. If a new or improved machine is to be made the basis of an entirely new manufacturing method, then 'the *whole* expense attending its operations should be fully considered before its construction is undertaken'. As a rule of thumb, the production of the first machine is about five times as expensive as the tenth.

Division of labour is also feasible in the inventive process: one person may have the idea, another may do the drawings, and a third may carry out the experiments. But, for Babbage, the most important step is the actual creation of new things. This does not mean that only a few people can be inventors; in fact, many have the ability to design mechanical devices, but lack the patience, will-power and application. It should be realized that 'the great success of those who have attained to eminence in such matters was almost entirely due to the unremitted perseverance with which they concentrated upon their successful inventions the skill and knowledge which years of study had matured'.[11]

4.3 INVENTIONS AND INVENTORS

Babbage makes a clear distinction between inventions and the application of inventions, a distinction that was introduced into economic theory by Schumpeter[12] around 1912. An invention is a new idea, an original way of utilizing physical and chemical phenomena, or a combination of technical operations already known into a new process. However, the application of inventions on a large scale may be prevented by technical or economic difficulties.

The practical significance of the distinction between inventions and innovations, to be discussed later, depends partly on whether technical development is seen as a continuous process of innovation or as a discontinuous process, with innovations in separate 'bursts'. Schumpeter views innovations as just such 'bursts' or 'shocks', which startle society when they occur from time to time. For him, idea and application are two basically different things. But if technical development is taken to be a more or less continuous stream of innovations, then there is a closer connection between inventions and their applications, and we are likely to regard neither of them as an entirely unexpected event but rather as a systematic development, inherent in the nature of things. In this case, the distinction between invention and innovation becomes rather blurred.[13]

However, the distinction is useful in helping us decide whether inventions are the more or less accidental result of personal inspiration or the product of technical, economic and social conditions. If technical development is viewed as a series of separate 'bursts', it is more convenient to regard inventions as arising from personal inspiration; if it is viewed as a continuous stream of improvement and innovation, this is more compatible with the deterministic notion that inventions are the result of the prevailing conditions. In the latter case, these conditions necessarily lead to innovations in a systematic manner in which one invention triggers off the next one. The inventor is usually anonymous, partly because similar inventions are being made elsewhere at the same time, and partly because the results are gradual step-like improvements.

One should be careful to avoid typecasting the inventor. Some believe inventors are geniuses; others that *all* engineers and technical people are inventors. Babbage, by the way, makes it clear that the remark, attributed to Edison, that an invention is ninety-nine per cent perspiration and one per cent inspiration, applies very much to himself. Inventions that can hardly be conditioned by historic necessity are at the same time rarely due to supernatural inspiration, and one can generally point to a set of factors on the basis of which the inventor arrives at a new concept by means of accurate observations and the combination of ideas. The important point is that the inventor has a creative mind which rejects the status quo and strives to modify it.

Not all technical people are creative, and presumably most engineers do not do much more than maintain the status quo. Some technical people who are creative are not regarded as inventors since their activities are only directed towards the regular and systematic improvement of production processes; however, their motivation springs from the same creative mentality as that of the inventor who—unlike them—steps out of anonymity. Technical people whose activities form part of continuous technical development tend to propose innovations that can be put into practice; their 'inventions' are mostly innovations, since these inventors take into account both the technical and the economic prerequisites of putting a suggestion into practice. This connection between invention and application is clearly looser when the activities of the inventor do not form part of a continuous stream of innovations but are instead the incidental results of his creative mentality.

4.4 THE STEAM-ENGINE AND THERMODYNAMICS

The steam-engine was developed mainly because human and animal labour had proved inadequate for pumping water from mines. Around 1700 Savery constructed a pumping system in which the necessary vacuum was created by condensing steam, generated by heating water. The same principle was used in Newcomen's steam-engine in 1712, which gave a considerably better performance because of its refined mechanical construction. This steam-engine consisted essentially of a vessel closed at the top by a piston, one end of which was connected to an arm. Steam was introduced at the bottom of the vessel, after which water was led in via another route. The condensation of the steam created a temporary vacuum which allowed the higher external pressure to push the piston down. This shifted the arm from the slanting rest position, so that the end not connected with the piston moved up. The arm then assumed the rest position once more, and a new cycle began with the introduction of more steam into the vessel, each movement of the arm representing a stroke of the pumping operation.

James Watt[14] (1736–1819) was an instrument-maker employed by Glasgow University, where in 1763 he was given one of Newcomen's 'atmospheric engines' to repair. Watt hit on the idea of separating the steam condensation from the rest of the process by carrying it out in a second vessel, a condenser. This presented various technical difficulties, solved only because sufficient money was put up by a manufacturer, Matthew Boulton (1728–1809), who took on the risk of working with Watt.[15]

After twenty years of hard work, Watt made a steam-engine in 1785, which differed from Newcomen's in the following three important respects. First, the cylinder or steam vessel was separate from the condenser, so that the temperature could be kept high in the first and low in the second. This raised the efficiency, because the cylinder did not have to be reheated after the condensation. Secondly, modification of the arm mechanism opened the way for other uses than in pumping water out of mines. The steam was introduced at alternate ends of the cylinder, so the piston, and thus also one of the ends of the arm, moved up and down. The coupling between the piston and the arm had to be adapted to the upward movement. Using a flywheel, Watt devised a mechanism that converted the piston's up-and-down movement into circular motion. Finally, Watt incorporated a 'governor' which automatically reduced the amount of steam when the engine was running fast enough. As Ferguson says,[16] Watt's steam-engine was in this respect the forerunner of modern 'feedback' systems.

The steam-engines built around 1800 differed from Watt's in that the steam pressure was now very high and the condenser had been dispensed with. This put great demands on the metal components and the joints, and there were many accidents. But development of the high-pressure steam-engine was continued, mainly because people realized its significance for the locomotive. An important step was Evans' idea of putting the cylinder and the crankshaft on one side of the arm, so that the size and weight of the arm could be reduced. In the 'compound' steam-engine, used in ships around 1860, the steam was led into several cylinders. The principle of Watt's steam-engine is that part of the heat introduced in the system is converted into work, the rest going to a section whose temperature is lower. The heat converts the water into steam in the cylinder at a high temperature; in the condenser the steam turns into water, and the remaining heat is liberated.

An interesting feature of the development of the steam-engine is that it was followed—not preceded—by the theoretical explanation of the phenomena involved. In this case science was stimulated by technology, and the debate on the efficiency of the steam-engine introduced a new branch of science—thermodynamics. This new science was mainly the work of Carnot,[17] who derived an expression for the efficiency of the steam in the following way. If Q_1 is the amount of heat introduced at a temperature T_1, if Q_2 is the amount of heat removed from the system at a temperature T_2, and if the amount of work done is denoted by A, then the efficiency ε is given by:

$$\varepsilon = \frac{A}{Q_1} = \frac{Q_1 - Q_2}{Q_1} = \frac{T_1 - T_2}{T_1}$$

This formula shows that the greater the temperature difference between the boiler and the condenser, the higher the efficiency of the steam-engine. There is therefore a limit to the efficiency of the conversion of heat into work in a cyclic process. This is expressed in Kelvin's formulation of the Second Law of Thermodynamics, according to which it is impossible to extract heat from a heat source and fully convert it into work without compensation. Thus, a ship cannot be propelled by converting the heat content of the sea into work. It is impossible for heat to flow spontaneously from a colder place to a warmer one.

On the other hand, the First Law of Thermodynamics deals with the conservation of energy and is implied by the equality between the work done, A, and the difference between the heat introduced and the heat extracted $(Q_1 - Q_2)$ in the above formula for the efficiency of a steam-engine. Treating heat as a form of energy, Joule determined the mechanical equivalent of heat in about 1840.

The introduction of the concept of entropy made it possible to predict the direction of spontaneous physical and chemical processes in nature. The entropy always increases in irreversible processes, such as the flow of heat from a warmer body to a colder one, occurring in a closed system, i.e. without external interference. An increase in entropy is synonymous with an increase in the degree of disorder. The extent to which energy is available for doing work is given by the free energy, which is a function of the internal energy and entropy. Thus, a sack of coal represents free energy that can be used for doing mechanical work, and when it is being used for this purpose, the free energy decreases and the entropy—or degree of disorder—increases.

These thermodynamic considerations indicate that heat cannot be converted into work efficiently, and it is better to convert free energy directly into mechanical or electrical energy. However, it will not be very long before the free-energy reserves, which solar energy has built up in the form of fossil fuels in the course of millions of years, are exhausyed. This run-down is irreversible, and is therefore different from the cyclic processes found in nature. The consequences of this for the problem of energy resources and the environment will be discussed later (see Section 12.6).

4.5 BESSEMER STEEL

The development of the iron and steel industry in Britain from 1780 to 1880 involved many technical changes of different calibres,[18] and once again practice preceded theoretical explanations of the reactions involved. Technical people with a practical bent experimented on a large scale. There were many disappointments, and the road from idea to application was long and arduous. This is clearly illustrated by the life of Sir Henry Bessemer (1813–1898), who, after having made a number of inventions, took his latest—an automatic gun—to France in 1853, where he was told that the iron of which the gun was made was not strong enough. From then on he devoted himself to metallurgy with the aim of making a stronger type of iron.

Until about 1850 industry produced mostly wrought-iron and a little steel. The latter was made by re-melting iron in the presence of carbon, which was a slow process. Bessemer took cast-iron containing about four per cent carbon, melted it and blew air through it. In this process carbon monoxide and carbon dioxide are formed, so that most of the carbon is removed from the metal in the form of these oxides; the silicon present is also bound. The heat generated by the reaction is sufficient to bring the temperature above the melting point of carbon-free iron, so that no external heating is needed for that purpose. Bessemer described his invention in a famous lecture given in 1856 before the British Association for the Advancement of Science. Several licences were taken out on his patent at once, but unfortunately his process did not seem to work for others. His star, which had risen so suddenly, was in danger of falling. It took years to discover why the cast-iron Bessemer used could be made into steel while other samples could not. It happened that Bessemer had used a batch of crude iron that had a very low phosphorus and sulphur content. This finding indicated that the difficulties caused by phosphorus could be overcome by using Swedish crude iron, which had a very low phosphorus content. But it was only after the discovery that manganese was necessary to remove oxygen and sulphur from the iron that the Bessemer process could be used for the large-scale steel production.

After the initial failures in 1856 it took a great deal of faith in the principle of the process to carry out experiments on a large scale. Courage and perseverance were essential ingredients of the attempts to improve the steel-making process in such a way that the price of the product would be low enough to guarantee a large market. Bessemer himself believed that his attempts would be successful:

I could now see in my mind's eye, at a glance, the great iron industry of the world crumbling away under the irresistible forces of the facts so recently elicited. In that one result the sentence had gone forth, and not all the talent accumulated in the last 150 years . . . no, nor all the millions that had been invested in carrying out the existing system of manufacture, with all its accompanying great resistance, could reverse that one great fact.[19]

After 1860 Bessemer steel spread rapidly as a substitute for iron, and was extensively used for the building of railways.[20]

4.6 ELECTRICITY AND EDISON

While the previous two examples have suggested that technology precedes science, it was different with electricity: 'If science owed anything to the steam-engine, it paid its debt to technology with electricity'.[21]

The term 'electrical force' was first used by William Gilbert (1544–1603), physician to Queen Elizabeth I, to describe the attraction observed between two bodies of different materials (for example a glass rod and a piece of silk) after they have been rubbed together. Scientists like Boyle, von Guericke and Musschenbroek isolated electrical charges and studied their effects. Benjamin Franklin realized that lightning was an electrical phenomenon and constructed the first lightning conductor.

The contact between science and practical innovation gave a new impetus to experiments. Galvani, an anatomist, discovered by chance the electrical effect of metals when he was dissecting frogs. The physicist Volta interpreted this around 1800 and constructed a battery from small containers filled with an acid and connected by silver and zinc plates, the device being capable of generating an electrical current. About twenty years later, Oersted accidentally discovered that a magnetic field is created if a current is passed through a wire. Ampère suspected that magnetism could be explained by the movement of electrical charges, wrote a treatise on the interaction between two currents, and showed that electromagnets were stronger than ordinary ones. Faraday (1791–1867) succeeded in generating a current in a conductor by the use of magnetic forces, thus laying the scientific foundations of the dynamo, the generator, the electric motor and the electric powerstation.

The other source of electricity, from chemical reactions, was used for the construction of simple telegraph systems. The improvement of communication systems was the aim of Edison, who invented the telegraph in 1870 for Wall Street; this device was capable of transmitting and printing three hundred letters a minute. This avenue of experiment reached its climax in the invention of the telephone in 1876.

Thomas Alva Edison (1847–1931) has numerous inventions to his credit, but he is perhaps best known for the invention of the lightbulb. Edison was concerned from the beginning not with an invention (i.e. a scientific principle) but with an innovation (i.e. a new, economically feasible product suitable for mass production). It was a special feature of his genius that he concentrated on developing commercially practicable processes. Almost all his inventions were creative syntheses of technical and commercial principles. His idea that electric lighting should be based on lamps connected in parallel is an example of 'economic reasoning, ahead of the best thinking in science of that day'.[22]

4.7 MARX REVISITED

We now return to Marx because *Das Kapital*[23] displays a thorough knowledge of the state of technology of the time. Following Ure, Marx distinguished between machines for generating power ('motor machines'), machines for transmitting power ('transmitting machines'), and machines for doing work ('working machines'), the last of which being the starting point of the Industrial Revolution in the eighteenth century. An increase in the capacity of the machines used for doing work called for more powerful prime movers. Marx mentions the genius of Watt, whom he quotes as saying in 1784 that his steam-engine should be regarded 'not as an invention for a specific purpose, but as an agent universally applicable to Mechanical Industry'. As the power-generating machine becomes more and more independent of human labour, it can drive several working machines at the same time. However, an individual machine can be replaced by a 'machinery system' only when 'the subject of labour goes through a connected series of detailed processes that are carried out by a chain of machines of various kinds, the one supplementing the other'.[24] The result is an automated system on a large scale. According to Marx, as the prime movers and the power-transmission devices become larger, industry undermines its own technical basis in so far as is the manufactory or workshop, which could not have produced the printing press, the mechanical loom and the carding machine. The revolution in the relations of production—the technical development—results from the development of technology.

Marx's views on the economic and social changes which lay ahead were based to a large extent on his accurate observation of technology in his own time, and of the changes in it. He distinguished between the physical and the 'moral' wear of machines, anticipating the distinction made in business economics between the technical and the economic life of machines. However, unlike traditional business economics, Marx's economics dealt with technical changes in which old machines are not replaced by identical ones.

4.8 CONCLUSIONS

Step by step man has extended his capabilities. While ideas may come in 'bursts', their realization is by and large a gradual process—a conclusion that will be examined further in the chapter on Schumpeter. Most inventions are associated with a name and are, in this sense, personal, but countless changes in production techniques are due to people who have remained anonymous. Yet these people have made a significant contribution to technical development.

An invention is a personal creation. The steam-engine would certainly look different if Watt had not existed and, without

Boulton, Watt would not have been able to develop it. But once the steam-engine was there, 'the process of replacement became irreversible'—as Ferguson said.[25] An invention is sometimes so timely that it can be utilized at once on a large scale; in other cases, such as Babbage's computer, it is so remote from the social and economic atmosphere of the time that it is a resounding failure.

While an invention is often completely novel, sometimes it consists of an elegant combination of elements that are already known. An invention is not always an innovation. Not all inventors have a dramatic invention like the light-bulb to their name. But they all have certain characteristic in common: they reject the technical status quo, they want to work things out for themselves and they have enough perseverance to succeed where others have failed. While the cost of original thought is just the cost of the thinker's effort, its actual realization often needs expensive equipment. Marx's characteristic remark in this connection is: 'Once discovered, the law of the deviation of the magnetic needle in the field of an electric current, or the law of the magnetization of iron, around which an electric current circulates, cost never a penny'.[26]

Technical development is not homogeneous, for it comprises both the epoch-making changes produced by the great inventors and the infinitesimal changes effected by men whose names are not remembered. These essential characteristics are lost in a purely quantitative treatment of technical development in terms of production and the factors of production.

Apart from Marx, economists in the first half of the nineteenth century were hardly interested in technology and its effect on society. In the next chapter we shall see if this situation improved subsequently.

5 The beginnings of production theory

In science and philosophy nothing must be held sacred. Truth itself is indeed sacred, but where is the absolute criterion of truth?

Jevons[1]

5.1 INTRODUCTION

Soon after the publication of volume I of *Das Kapital* economics began to concentrate on the subjective theory of value, treating prices as if they were almost entirely the result of the subjective value that consumers put on goods, so that the behaviour of producers became to some extent a reaction to that of consumers. Although the theory of consumer behaviour was rigorously developed, the analysis of production did not by any means stagnate. In fact the period 1870–1920 saw the foundation of the neo-classical theory of production and the frequent use of the Cobb-Douglas production function. Following in the footsteps of the Classical economists, authors like Wicksell and Wicksteed gave a great deal of thought to production, and the capitalist system of production came under the scrutiny of the Austrian school, and in particular Böhm-Bawerk. While the theory of value and price received a sudden impetus in 1870, the development of economic thought on technology was a more gradual process, as we shall see in this chapter. We shall start with Cournot, since economists with a pronounced subjectivistic bent, like Walras[2] (1801–66) and Gossen[3] (1810–58), did not deal with technical change.

5.2 COURNOT

Cournot (1801–77) was a French mathematician and philosopher, and his contribution to economics was more a digression from his subject than a logical extension of it. But this does not diminish its importance for modern economic theory. We do not know what prompted this 'digression', and his *Souvenirs*,[4] written in 1858, give no clue. It seems likely that, having some knowledge of the work of Adam Smith, Ricardo and Say, and specializing in the theory of functions and in calculus, he realized that economics could become a more exact science with the use of mathematical functions.

His main contribution to economics lies in the introduction of the demand function[5] and in the accurate calculation of this for situations in which suppliers range from a single one to an infinite number. In retrospect the calculation of the market equilibrium in the case of monopoly and duopoly seems an incredible achievement, but for Cournot it was simply the application of the hypothesis that 'everybody tries to derive the greatest benefit from his work and possessions'. Cournot concentrated mainly on the demand side of the economic process and did not formulate a production function, although in his book published in 1838 he introduced a total cost function and recognized the importance of the marginal cost function for the 'solution of the main problems of economic science'.[6] The absence of a production function was one of the reasons why technical change had no place in Cournot's mathematical analysis.

But technical change is discussed in the non-mathematical editions of Cournot's book which were published in 1863 and 1877 (receiving hardly any more attention than his earlier work).[7] In the 1863 edition he deals with general aspects of 'progress', and in the 1877 edition, published shortly before his death, he analyses the industrial production process by a comparison of machines and factories. He then discusses the social effects of 'industrial progress' on the labour market, concluding that there is no problem as long as technical change raises the quality and quantity of goods and lowers price without causing a decrease in the demand for labour, but when it does cause a permanent drop in this demand we are forced to make a value judgment about the ultimate aim of economic activity. He believes there are no objective criteria for deciding the direction of economic trends, 'since we would have to measure and compare items that are neither measurable nor comparable'. The choice of money as a yardstick for economic progress cannot be maintained in the face of 'objective criticism'.[8] Cournot clearly understood the difficulty of the welfare evaluation of technical change, but his analysis of the latter does not come up to the level of his contribution to the theory of prices. It is interesting to note that Cournot's remarks on technical change have been further developed by the famous Soviet economist, Dimitriev, who can also be considered Sraffa's forerunner.[9]

5.3 JOHN RAE

The Scotsman John Rae (1796–1872) published a book[10] in 1834 which according to Schumpeter[11] 'far surpassed the economists who were successful' both in its originality and in its breadth of vision. Rae's importance for the development of the theory of capital and interest was later recognized by Böhm-Bawerk[12] and Fisher.[13] He introduced the concepts of time preference, roundabout production method and liquidity preference, and analysed the connections between them.[14] One can hardly overestimate Rae's importance as a forerunner of both Böhm-Bawerk's and Fisher's theories of capital and interest, but Rae contributed even more to the economic theory of technical change.

Unlike Adam Smith, who thought the interests of the individual coincided with those of society, Rae thought the promotion of individual welfare could not be identified with the interests of society as a whole, this lack of coincidence being closely connected with the creative nature of technical change. Technical development is a major means of increasing the riches of the society, because this process is 'a creation, not an acquisition', while it plays no part in increasing the riches of the individual because this 'may be simply an acquisition not a creation'. The objectives of society are different from those of the individual, the former pursuing innovations, the latter possessions. Their means of achieving these are also different: work and saving raise the assets of the individual, while only the 'inventive faculty' can raise the assets of the society. Unlike Adam Smith, Rae therefore believes that it is an important function of government to promote scientific and technical development and encourage inventions and improvements in existing techniques.

Anticipating the recent discussions on the subject of technical change embodied in capital goods, Rae says 'before capital can increase, there must be something in which it may be embodied'. Only technical development brings profit to capital formation and provides the population with sufficient goods, so that it has 'most title to be ranked as the true generator of states and people'. Rae in effect adds the application and diffusion of new technology to the accumulation of capital and the division of labour as the main factors contributing to the prosperity of society.

Rae devotes a complete chapter to technical change and its consequences. He first considers the psychology of the inventor and concludes that he is motivated by the same urge to create as the artist or the mathematician; they are all people who do not imitate but create new things, whether in the spiritual, mental or material field. The handing-down of existing knowledge is still important, but those who do nothing else 'neither oppose, nor direct the current'. However, technical development depends not only on inventors but also on natural resources and the discovery of the laws of nature:

The progress of knowledge of the nature and qualities of particular substances, gradually introduced a knowledge of the properties and natures of substances in general. Men first see in the concrete, afterwards in the abstract. Thus, the discovery of the several mechanical powers, and the knowledge acquired of the nature of each, led in time to the general principles of mechanics.

Rae distinguishes between accumulation of capital (an increase in the stock of capital goods without a qualitative change in it) and the augmentation of capital (an increase in the stock of capital goods with a concomitant qualitative change in it). With mere accumulation of capital, the profits diminish, whereas when technical development is embodied in new capital goods the profits increase.[15]

Rae, a somewhat neglected figure whose importance is now being recognized, was not only a pioneer of the theory of capital but also made a significant contribution to the theory of technical change.

5.4 MENGER, JEVONS AND WALRAS

There is an understandable tendency to regard 1870 as a turning point in the development of economic theory, because the publications of Menger, Jevons and Walras appeared around that time. Although there was a new emphasis on the behaviour of the individual consumer, as a reaction to the neglect of Classical theory, the marginal utility school did not completely ignore the objective aspects of production: nearly all of them paid some attention to technology and technical change, and Böhm-Bawerk and Wicksell, in particular, produced an important analysis of the effect of capital formation on economic development.

Menger (1840–1921), the founder of the Austrian school, did not give much thought to technology and when he did deal with it he concentrated on the process of the creation of new products and the improvement of old ones. In his system the structure of production is determined mainly by the subjective value the consumers put on 'goods of the first order' and not by the underlying economic framework. Menger does not mention technical development amongst the causes of increasing prosperity.[16]

But although the Austrian economists are regarded as exponents of a theory of static equilibrium, they did not ignore technical development altogether. In an account of capital formation, Menger says 'human life is a process'[17] and establishes a connection between the productivity of capital and the systematic underestimation of future needs. His terms of reference are utility, benefit, and satisfaction of needs, but by treating capital as a factor of production he introduces an element which is to a great extent determined by the technology used in the production process. Technical development—as far as one can perceive it in Menger's work—is the acquisition of knowledge about new products.[18]

Jevons (1835–82) paid considerable attention to capital formation although the subject of his book[19] gave him little opportunity to discuss actual economic development in the tradition of the British Classical economists. Jevons thought that capital can open up roundabout routes of production; in this respect, he anticipated Böhm-Bawerk and Wicksell, especially since he viewed production time as a variable factor.[20] He pointed out that 'the amount of capital invested' and 'the amount of investment of capital'[21] should be clearly distinguished. If p is the capital invested and t is the production time, then pt is the investment, and if the production cycle is repeated, the total investment is given by Σpt. If regular investment and disinvestment are in progress during the period t, the investment is given by $\frac{1}{2}pt$, if p is the largest capital invested. Jevons based the demand for capital on the marginal productivity of capital.[22]

In volume two of his *Eléments d'économie politique pure*,[23] published in 1877 (and substantially based on an article[24] written in 1876), Walras (1834–1910) devotes a chapter to capital. The demand for and the supply of capital are treated in monetary terms and are incorporated in the general equilibrium system. But the formal approach vanishes when Walras discusses the idea of progress in an economic system of perfect competition.[25] He makes an important distinction between 'economic progress' and 'technical progress'; in the former the technical coefficients are changed quantitatively by the use of more capital and less land, while in the latter there is a qualitative change in the technical coefficients, for some of them disappear and are replaced by new ones.[26] This distinction will be discussed later in connection with the production function (see Section 8.8).

Since Menger, Jevons and Walras placed considerable emphasis on capital formation, they did not really break the continuity of the Classical tradition. Rather, they paved the way for an elaborate theory of capital and, thus, of economic development.

5.5 BÖHM-BAWERK

Schumpeter called Böhm-Bawerk (1851–1914) the 'bourgeois Marx'[27] because of his detailed analysis of the capitalistic system. In the present context, however, he forms a bridge between Classical and modern economists. The 'subsistence fund' of his capital theory is similar to the Classical wage fund. The subsistence fund makes it possible to follow roundabout routes of production, and the larger the fund the longer these routes can be.[28] If well-chosen roundabout production methods are followed with the aid of capital goods, more and/or better consumer goods can be obtained than by direct production. But the longer this roundabout route of production, the slower the growth of production. The consumer goods produced at the end of such a route must be set against the disadvantage of lower consumption for the duration of this longer production period, which Böhm-Bawerk called a 'sacrifice to time'.[29] But the more extensively the durable consumer goods are made by longer production methods that are qualitatively better, the more the rise in the utility outstrips the rise in the production cost.[30]

We now examine some controversial features of Böhm-Bawerk's theory of roundabout production methods.[31] First of all he makes it clear that both increased productivity and technical progress are possible without an extension of roundabout production methods, for an invention often leads to a less roundabout production method. There is no consistent relationship involved, although Böhm-Bawerk believes that inventions associated with a more roundabout production method form the majority. Inventions which shorten this route quickly oust existing methods, whereas the realization of the opposite type of invention depends on the relatively slow growth of the subsistence fund. Since the anticipated profit determines whether a roundabout production method is applied or not, both recent and older inventions will be in use at any given moment. More technical knowledge exists than can be put to use, due to scarcity of capital. The gap between these two levels of knowledge is due to production methods that follow a more roundabout route than the methods that are actually in use, since 'all the shorter ones, which are at the same time technically more advantageous, have already found application and achieved prominence, even without an increase in the material funds'.[32]

Böhm-Bawerk distinguishes between a lengthening of the roundabout route of production at a given level of technical knowledge and the influence of inventions on the length of this route.[33] The introduction of the concept of the production period follows from Böhm-Bawerk's starting point, which is that only the original factors of production—labour and scarce natural resources—are important for characterizing capitalistic production. Of the two types of capital, only the circulating type is involved. The concept of the roundabout production method and the related concept of a production period only adequately fit into a model in which circulating capital is the only type taken into account; when fixed capital is introduced as a factor of production these two concepts can no longer be used as analytical tools,[34] because the length of the production period cannot then be accurately defined. These concepts may have limited validity, but Böhm-Bawerk's basic idea—the recognition of what may be called the intertemporal nature of production—was important,[35] and intertemporal production functions will be discussed later in connection with microeconomic production theory (see Section 8.4).

Böhm-Bawerk gave only a non-mathematical description of this theory. Wicksell,[36] however, produced an elegant mathematical summary of its essence in a brilliant publication appearing in 1893. Wicksell's presentation can be translated directly into a model in which labour is the only factor of production. We assume a static situation, involving a given number L of employees, an annual per-capita production of q, an annual wage per head of w, a length of the roundabout production route (i.e. the production time) of t, and a circulating capital of K (which is invested for an average time of $\frac{1}{2}t$). Böhm-Bawerk's theory can be reduced to a relationship between q and t. He assumes that q increases with increases in t, but that the marginal returns decrease. We thus have the function $q = f(t)$; if total production is Y, we can also write $Y = q.L$. Recalling Jevons, we can easily see that the amount of capital invested[37] will be given by $K = \frac{1}{2}t.w.L$, and the total production Y is also equal to the total income, so that $Y = w.L + r.K$, where r is the rate of interest or profit.

Wicksell discusses two cases, in which the length of the roundabout production route t is such that the annual wages are maximized (first case) and where the rate of profit r is maximized (second case), the latter conforming to Böhm-Bawerk's theory, as Wicksell himself remarks. From the relationships

$$Y = q.L$$
$$K = \tfrac{1}{2}t.w.L$$
$$Y = w.L + r.K$$

it follows that

$$q = w + \tfrac{1}{2}t.w.r$$

Writing r as a function of t and assuming that the wages w are given (so that $w = \bar{w}$), this leads to:

$$r = \frac{q - \bar{w}}{\frac{1}{2}\bar{w}t} = \frac{f(t) - \bar{w}}{\frac{1}{2}\bar{w}t}$$

It is now possible to find the length of the roundabout production route at which the rate of profit r is maximized. A graphical means of finding the best production period[38] is shown in Figure 4, where the length of the production route t is plotted along the horizontal axis and the per-capita production q along the vertical axis. The given wage $\bar{w}$ is OB, and the optimum production period is OA. In the case of a higher wage, this period will be longer.

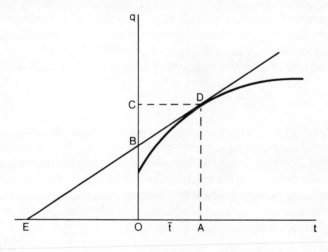

Figure 4

As will be seen in Chapter 9, we can use this graph to demonstrate the relationship between Böhm-Bawerk's and Wicksell's production period and Harrod's interpretation of neutral technical change. It is not entirely clear whether technical change is represented in Böhm-Bawerk's treatment only by a displacement along the graph of the intertemporal production $q = f(t)$ or also by a shift in the whole graph. Nevertheless, refinements in the modern analysis of production and technical change owe an unmistakable debt to his work.[39]

5.6 VON THÜNEN

It is rather surprising that Böhm-Bawerk did not formulate a production function in view of his explicit references to von Thünen[40] (1783–1850), a shrewd German landowner who spent the winter months on his estate in Mecklenburg analysing economic problems[41] and whose true contribution to this subject was recognized in the last century only by Marshall.[42]

In the second edition of his work, von Thünen describes the intensive variant of diminishing returns and considers production the result of two factors of production: labour and capital. He explicitly formulates the concepts of the marginal products of labour and capital and he equates the wage rate with the value of the marginal products. But the most remarkable thing is that he added a mathematical formulation to his description of the concept of the production function ('after more than twenty years of fruitless efforts') and worked out several numerical examples, using empirical data from his estate. His general production function is of the following form:

$$p = h(g + k)^n \tag{1}$$

where h, g and n are constants, p is per-capita production, and k is the ratio between capital and labour.[43]

Von Thünen is to modern production theory what Cournot is to the theory of monopoly and duopoly; to demonstrate this, we assume the special case of $g = 0$, denote the total number of employees by L, total production by P, and capital by K. Equation 1 can then be written as:

$$P = h.L(K/L)^n \tag{2}$$

and this can in turn be written as:

$$P = h.L^{1-n}.K^n \tag{3}$$

If we consider that von Thünen always chose values between 0 and 1 for n, we find that he virtually formulated the famous Cobb-Douglas production function.[44] Von Thünen's starting point in describing the technique of the production process was that 'part of the workers can be replaced by capital, and—conversely— part of the capital can be replaced by extra workers'.[45]

5.7 WICKSTEED

In a remarkable study published in 1894, Wicksteed (1844–1927) developed further Jevons' vague notion of a production function.[46] Admittedly, Edgeworth[47] introduced a production function in 1893 in a footnote to a discussion of von Wieser's book, but Wicksteed was the first to develop this concept in clear and precise terms.[48]

Wicksteed introduced the production function in the context of the theory of marginal productivity. It is interesting that Wicksteed first proposed a general-welfare function, where the 'total satisfaction (S) of a community is a function (F) of the commodities, services, etc. ($A, B, C, \ldots$)'. If the amount of a good is denoted by K, then ds/dK gives the 'marginal efficiency or significance of K as the producer of satisfaction'. The exchange value of the whole stock is then

$$\frac{dS}{dK} K.$$

After giving this general definition, Wicksteed proceeds to treat social welfare as a product, and the goods and services as factors of this product: 'Let the social product (P) to be distributed, be regarded as a function (F) of the various factors of production ($A, B, C, \ldots$). Then the (marginal) significance of each factor is determined by the effect upon the product of a small increment in that factor, all the others remaining constant'. Simultaneously with the introduction of the production function, Wicksteed thus laid the foundation of the theory of marginal productivity.

Wicksteed works with a production function $P = F(L,K)$ such that $\lambda P = F(\lambda L, \lambda K)$. He bases this property of the production function first on its invariable nature, then on the proportional application of labour and capital to a given piece of homogeneous land, so that a constant yield per unit area is produced, and finally on the recurrence, without any change, of the 'physical conditions under which a certain amount of wheat, or anything else, is produced'.

Wicksteed does not mention that the function F must be linearly homogeneous, and in fact—as Barone has pointed out—he gives the impression of deriving $2F(L,K) = F(2L,2K)$ from $F(L,K) + F(L,K) = 2F(L,K)$. Wicksteed also assumes that it is justifiable to 'express a perfectly heterogeneous aggregate of factors in terms of a single unit'. He thinks that this thorny problem is solved by the equivalence between the marginal product and the marginal cost of the factor of production under equilibrium conditions.[49]

These then are the main features of Wicksteed's study, a testimony to the power of independent, original thinking, which is scarcely weakened by some mathematical errors. Later Wicksteed withdrew the gist of his argument,[50] because—it now seems[51]—Pareto had said that the determination of the marginal products as partial derivatives was inadmissible. In his book published in 1910 Wicksteed completely abandons his formal analysis and more or less follows Marshall's footsteps in giving a number of empirical characteristics of production. He distinguishes between laws of returns according to the number of factors of production regarded as variable. Although the actual state of technology is described more accurately, less attention is paid to the theory of technical change.

5.8 BARONE

Enrico Barone (1859–1924) was an Italian army officer who made some important contributions to mathematical economics. In a letter to Walras, dated 20 September 1894, he said he had succeeded in formulating a simple theory of distribution 'on the basis of a new theorem of a maximum net profit for the entrepreneur'.[52] He uses symbols to denote the size both of output and the factors of production. His starting point is that the production function is *not* homogeneously linear, and—what is more remarkable—he formulates virtually the whole of the marginal productivity theory in a few sentences.

Barone returns to the linear homogeneity of the production function in an unpublished review of Wicksteed's book, discussed before. The original has been lost and the text exists only in Walras's translation. In this review, Barone exposes the flaw in Wicksteed's argument, which is that in his treatment of increases in scale Wicksteed attributes to the production function some characteristics it possesses only if it is homogeneously linear. In his formulation of marginal productivity theory, Barone makes a clearer distinction than

Wicksteed between actual and optimal conditions of production. Barone also shows that the essence of the marginal productivity theory lies in Walras's set of equations. In a letter dated 13 November 1895, Barone impresses upon Walras that the technical coefficients can also be regarded as variables, whose magnitude can be determined by the method of general equilibrium.

Discussing Wicksteed's theory in a study written in 1894 and 1895, Walras[53] took into account several of Barone's views; his study shows that the production function is indeed implied in his—i.e. Walras's—ideas of economic relationships, though he does not single out the relationship between production and the factors of production. In this respect Wicksteed and Barone delve deeper. Barone later developed the production theory further[54] in the framework of a partial equilibrium analysis, where he noted the phenomena of technical change but failed to fit it into his analytical system.

5.9 MARSHALL

Meanwhile Marshall's *Principles of Economics*[55] had appeared, in 1890. In his inaugural lecture at Cambridge in 1885, Marshall said that, for him, economic theory was not 'a body of concrete truth, but an engine for the discovery of concrete truth'.[56] While the *Principles* often mentions concrete economic phenomena, the formal analysis of them is sometimes weak and confused. But although less advanced than his contemporaries in the formal treatment of the production function, Marshall discusses in greater detail the limited significance of this function for economic growth.

Marshall concludes his discussion of the effect of mechanization on the division of labour by saying that 'any manufacturing operation that can be reduced to uniformity . . . is sure to be taken over sooner or later by machinery'.[57] Like Marx, Marshall displays a wide knowledge of technology, which enabled him to identify and project future developments. Thus, he mentions the elimination of human control work by the introduction of 'an automatic movement, which brings the machine to a stop the instant anything goes wrong'.[58] He believes that an improvement in machinery goes hand in hand with an increase in the division of labour, though the connection between them is not as close as many economists imagine. He also stresses that machines can work with greater precision[59] than the human hand, and realizes that this encourages a system of replaceable or interchangeable parts made by machines. He draws general conclusions from his observations of the textile industry and discusses how machines make work lighter.

Marshall points out the importance of new and cheaper means of communications and stresses the significance of the application and export of new inventions: 'English mechanics have taught people in almost every part of the world how to use English machinery'. Discussing the effect of mechanization on the ratio between agricultural and industrial activity, Marshall uses numerical data to show that not all those who have become redundant in agriculture find employment in industry because of the 'wonderful increase in recent years of the power of machinery', which permits a great increase in production to be achieved without a proportional increase in the labour force, so that the redundant labour drifts into sectors where 'the improvements of machinery help us but little'. It can be safely assumed that Marshall is thinking here of service industries, although he does not say so explicitly. This is borne out by his paradoxical statement that 'the efficiency of machinery has prevented the industries localized in England from becoming exclusively mechanical as they otherwise would'.

Starting with Babbage's preoccupations, Marshall discusses in detail the advantages of large-scale production, notably the use of specialized machines, the greater chance of inventing things, the financial resources for making mechanical equipment, more favourable purchasing conditions and better utilization of scientific and technical knowledge; 'changes in manufacture depend less on mere rules of thumb and more on broad developments of scientific principles'.

In connection with increases in scale, Marshall distinguishes between 'external economies' (derived from the general development of a whole sector) and 'internal economies' (derived from increased production within the enterprise). As well as the law of diminishing returns for agriculture, he believes in a law of increasing returns for industry, which is mainly due to better organization, in its turn due to larger inputs of labour and capital. He mentions, in passing, that one external diseconomy is 'the growing difficulty of finding solitude and quiet and even fresh air'.[60]

Unlike Marshall, Wicksell stresses the need to differentiate between (a) diminishing returns which arise when one of the factors of production is kept constant and (b) increasing returns, resulting from an increase in the scale of production when all the factors of production are variable. If 'large-scale and small-scale production are equally productive',[61] the production function is linearly homogeneous. As an example Wicksell produces a production function $P = a^{\alpha}.b^{\beta}$—with constant returns to scale $(\alpha + \beta = 1)$—where P is the production, and a and b are the quantities of the two factors of production. After Wicksell's death this relationship reappeared in the form of the Cobb-Douglas production function.

After his incidental introduction of a production function in 1893[62] Edgeworth wrote an article in the *Economic Journal* in 1911 in which he gave a broader analysis of related topics. Here he clarified the term 'diminishing returns' by distinguishing between diminishing marginal returns and diminishing average returns. He then showed that the shape of the curves for the returns is not altered when prices are introduced, provided that 'the prices, both of the product and the factors of production, remain constant while the amounts are varied'.[63] Finally he discussed the relationship between the shape of the curves for the returns and that of curves for the costs, distinguishing between the short-term and long-term, according to the factors in the production function kept constant.

5.10 CONCLUSIONS

Both the concept of the production function and its mathematical expression developed slowly. Since the second half of the last century the analysis has been closely linked with the distribution of aggregate production. This is especially true of those authors who had a fairly clear idea of the concept of a production function, namely Wicksteed, Barone and Wicksell.

But there was a price to pay for the clearer expression of the relationship between production and the factors of production: as the rambling, disjointed passages in the older literature were being disentangled and their gist extracted, some of the caution and qualifications were lost. Wicksteed's analysis of the production function is distinctly more elegant and incisive than Marshall's, but Marshall had a wider knowledge of actual production techniques.

As the importance of the production function increased, so the question of technical change receded into the background. Those who produced the most advanced mathematical treatment of the production function, i.e. Walras, Wicksteed and Barone, tended to ignore the changes caused by technology. In contrast, those with a less mathematical bent, like Böhm-Bawerk, gave more emphasis to the role played by technical change. The renewed interest in Böhm-Bawerk's 'temporal' theory of capital, in which time is a significant factor, may be due to the realization that the choice between various production methods that differ in their roundabout production route is tantamount to the choice between production processes that differ in their technical specification. If the production function expresses a dynamic relationship that includes technical change, we need to have a clear idea about the various types of technical change before we can incorporate this phenomenon into the production function. Wicksell, who has already emerged as a pioneer, did in fact classify various types of technical change after formulating his production function, as will be seen in connection with the macroeconomic production function in Chapter 9. However, we shall first examine, in the next three chapters, Schumpeter's theory, technical development in the first half of this century, and microeconomic production theory.

6 Schumpeter

The trouble with Schumpeter's theory is that it is descriptive rather than analytical.

Kaldor[1]

6.1 INTRODUCTION

Schumpeter (1883–1950) is often mentioned in the same breath as Marx because they both believed that 'changes in methods of production are a basic feature of capitalism'.[2] We saw in Chapter 3 that technical change—of a certain type—plays an important part in Marx's theory, and we must now examine the significance Schumpeter attaches to technical change.

Reading Schumpeter one realizes that the forte of this great economist lies not so much in a mathematical and analytical approach as in a historical description and a verbal theoretical analysis. Samuelson says that Schumpeter himself 'cheerfully admitted the difficulties he had in mastering and retaining mathematical techniques'.[3] Although, as Schumpeter himself remarked later, the concept of the production function 'had established itself as a result of the efforts of many minds'[4] around 1900, he made very sparing use of it in his own work. His inability to put his ideas about the development of economic life into a mathematical form may eventually change our assessment of him. But whatever the final evaluation of Schumpeter may be, it cannot be denied that he gave a new direction to the development of economic science by posing some entirely new questions. Schumpeter's preoccupation with the dynamics of economic life broke the spell of the static approach to economic problems.

6.2 HALF A CENTURY OF SCHUMPETER

In his first important work, written at a very early age in 1908, Schumpeter devoted some thought to the development of economic life,[5] which he called 'dynamics', and which was always to preoccupy him.

Before outlining some dynamic problems, Schumpeter makes some remarks on the place of technology and technical development in economics. Although he expressly states that 'the facts of technology influence the course of economic life, and the laws of the latter influence progress and the practical application of technical methods', he believes that attention should be given first of all to the factors that influence the application of new techniques. Apart from more or less fortuitous factors, such as the extent of the knowledge of new technical methods, we are faced here with the state of technology and economic considerations, and Schumpeter believed that economic science can take a big leap forward by dealing with technical phenomena.

The importance Schumpeter assigns to the creation of money in this early book also foreshadows his later work. He believed that since in static analysis there is no room for (a) new combinations of factors of production, (b) the entrepreneur (in the Schumpeterian sense), and (c) profit, so there is no need for the further creation of money. He also questions whether one can speak of economic development. Here he is not questioning that economic phenomena and magnitudes change, but suggesting that the cause of changes may not lie in 'economic factors'. This would be expressed today by asking whether economic development is explained by endogenous or exogenous factors. If the latter, Schumpeter would not speak of economic development at all, for he regards factors like population growth, consumer preferences, technical development and the social organization as non-economic factors. On the other hand, he says that changes in human nature and social organization can in fact be attributed to economic causes, which can be regarded in this connection as endogenous factors. Furthermore, every non-economic development (i.e. every development that cannot be explained by endogenous factors) has economic consequences, and he finally asks whether it is possible to formulate a theory of economic development based on the assumption of a fixed and immutable state of affairs.

The somewhat ambiguous terminology Schumpeter uses here for describing and explaining the development of the economic process is repeated in its entirety in his book on economic development, first published in 1912, then in a revised form in 1926 and finally in English in 1934. This is illustrated by the following passage from the 1934 edition:

By 'development', therefore, we shall understand only such changes in economic life as are not forced upon it from without but arise by its own initiative, from within. Should it turn out that there are no such changes arising in the economic sphere itself, and that the phenomena that we call economic development is in practice simply founded upon the fact that the data change and that the economy continuously adapts itself to them, then we should say that there is no economic development. By this we should mean that economic development is not a phenomenon to be explained economically, but that the economy, in itself without development, is dragged along by the changes in the surrounding world, that the causes and hence the explanation of the development must be sought outside the group of facts which are described by economic theory. Nor will the mere growth of the economy, as shown by the growth of population and wealth, be designated as a process of development. For it calls forth no qualitatively new phenomena, but only processes of adaptation of the same kind as the changes in the natural data. Since we wish to direct our attention to other phenomena, we shall regard such increases as changes in data.

We see that saving is no longer regarded as a factor leading to economic development in Schumpeter's use of the term. Capital formation—a consequence of saving—is a factor of growth, as is the increase of population. The terms 'development' and 'economic development' serve here not so much to describe the actual course of economic events as to distinguish the changes caused in the economic process by endogenous factors from other changes.

Not all endogenous factors lead to economic development, for Schumpeter explicitly excludes continuous endogenous changes. His theory of economic development is reduced to the treatment of spontaneous and discontinuous changes in the economic cycle. The endogenous changes Schumpeter has in mind are not found on the consumption side of the economic process, but on the supply side. Economic development consists of the discontinuous introduction of new combinations of products and means of production. The five examples mentioned by Schumpeter show that the term 'new combination' must be taken in a very broad sense; it comprises a new product, a new method of production, the opening-up of a new market, the utilization of new raw materials, and the reorganization of a sector of the economy.

In Schumpeter's view, new combinations can be financed only if a successful appeal can be made to banks to create money. He restricts the meaning of the word 'enterprise' to the creation of new combinations, and the meaning of the word 'entrepreneurs' to those economic figures who introduce new combinations. Schumpeter's entrepreneur operates in an uncertain world, has the courage to start up new ventures, and must be strong enough to swim against the tide of his society. Entrepreneurs are not inventors, since 'the innovations which it is the function of entrepreneurs to carry out need not necessarily be any inventions at all'.

According to Schumpeter, ups and downs in economic development can be explained quite simply by the fact that new combinations or innovations 'appear, if at all, discontinuously in groups or swarms'. The appearance of entrepreneurs in 'bursts' should be attributed exclusively to the fact that 'the appearance of one or a few entrepreneurs facilitates the appearance of others', which is the only reason for a boom.[6]

In his book on trade cycles, Schumpeter sharpened his analytical tools. He introduced the concept of the production function, which—in his words—tells us 'all we need to know . . . about the technological processes of production'. The production function is insensitive to small and permanent changes in production methods, but the introduction of new combinations should be regarded as the 'setting up of a new production function'. The changes caused by innovations are no longer regarded as economic development but as economic evolution. He uses the term 'technical development' only for innovations that involve the introduction of new methods of production. Generally, innovations should be distinguished from inventions. The application of new combinations by entrepreneurs is possible without inventions, while inventions as such need not necessarily lead to innovations and need not have any economic consequences.[7] Innovation itself is the independent endogenous factor that causes economic life to go through repeated cycles.

We now turn to Schumpeter's book on the development of capitalism, probable his best-known work. In his books mentioned so far he discusses growth, development, economic development and economic evolution primarily in connection with the causes of change, but in this book he also deals with the consequences of the evolution of capitalism. However, as the following quotation shows, his characteristic approach, in which he combines analysis with the definition of concepts, is still very much in evidence:

Capitalism, then, is by nature a form or method of economic change and not only never is but never can be stationary. And this evolutionary character of the capitalist process is not merely due to the fact that economic life goes on in a social and natural environment which changes and by its change alters the data of economic action; this fact is important and these changes (wars, revolutions and so on) often condition industrial change, but they are not its prime movers. Nor is this evolutionary character due to a quasi-automatic increase in population and capital or to the vagaries of monetary systems of which exactly the same thing holds true. The fundamental impulse that sets and keeps the capitalist engine in motion comes from the new consumers' goods, the new methods of production or transportation, the new markets, the new forms of industrial organization that capitalist enterprise creates.

The main characterizing feature of capitalism is the introduction of new combinations. Innovation, which earlier formed the basis of the concepts of development, economic development and economic evolution is now made identical with capitalism. This endogenous process of what Schumpeter calls creative annihilation is the 'essential fact about capitalism'. The large monopolistic enterprises effect the innovations and explain the special *modus operandi* of capitalism, manifested in the expansion of production. He admits that in principle the productive capacity of capitalism could also be explained by factors that are independent of capitalism. After rejecting several external factors, however, he asks if the special features of capitalism are the result of technical development, i.e. of the 'stream of inventions that revolutionized the technique of production'. He believes not, but rather that invention itself is a function of the capitalistic mode of production, so that technical development is in fact a consequence of capitalism—in his sense of the word—rather than an independent cause of the expansion of production.

The rate of growth of production is not reduced because the technical possibilities are exhausted, but capitalism suffers from a change in the position of entrepreneurs. On the one hand, it is now easier than before 'to do things that lie outside familiar routine—innovation itself is being reduced to routine. Technological progress is increasingly becoming the business of teams of trained specialists who turn out what is required and make it work in predictable ways', and on the other hand, characteristics such as personality and will-power 'must count for less in environments which have become accustomed to economic change . . . and which, instead of resisting, accept it as a matter of course'.[8] Economic development acquires an impersonal and mechanical character. The success of the capitalistic production makes capitalism itself redundant; the road to socialism is paved not by intellectuals and agitators but by large enterprises. Schumpeter's prognosis for capitalism is thus the same as Marx's, but based on very different reasoning.

Finally we should mention Schumpeter's idea of the production function in his monumental work on the history of analytical economics. His interpretation of the production function is clearer in this inexhaustible source of ideas than in his other works. In his first definition, the production function comprises 'the given technological possibilities *within the horizon* of producers'. A change in the technical horizon by the discovery of a new method of production or the making of a known process financially feasible leads to a new production function. A few pages later, Schumpeter, without realizing it, gives a wider definition of the production function when he places it 'in a world of blueprints, where every element that is technologically variable at all can be changed at will, without any loss of time, and without any expense'. Furthermore, he recognizes explicitly a third interpretation of the production function, which concerns enterprises that actually exist: a 'realistic' production function, which can be constructed on the basis of factual observations of production and factors of production, and which should be clearly distinguished from the 'logically pure' production function.[9]

6.3 ECONOMIC AND TECHNICAL DEVELOPMENT IN SCHUMPETER'S WORK

Hennipman rightly remarks that, almost from the very beginning, Schumpeter holds a certain point of view, which he basically retains throughout,[10] and which boils down to the application of new combinations in 'bursts' by men who then— and only then—can be called entrepreneurs. He uses the terms development, economic development, economic evolution and capitalism to denote a process of internal, endogenous changes, the essence of which is the introduction of innovations. This fusion between analysis and the formulation of concepts raises the question to what extent we can justifiably say that Schumpeter had a theory of economic development, a theory of trade cycles, and a theory of capitalism.

One might expect Schumpeter to elucidate these terms by telling us what factors influence them; but he does not do this. The concepts of economic development, entrepreneur and capitalism are insubstantial if his notion of innovations lacks an empirical content, and his theory of trade cycles is untenable if innovations occur not in bursts but in a continuous stream. Schumpeter tries to demonstrate with historical examples that important innovations occur discontinuously and cause 'waves' of different lengths.

As far as Schumpeter's theory of development is concerned, Higgins concludes that it has a 'large element of tautology in it',[11] although Kuznets considers the suggestion of a tautology unacceptable 'as a significant interpretation or extension of Professor Schumpeter's position'.[12] However, since the concepts are always described in terms of the application of new combinations, it is impossible to see in Schumpeter's ideas a theory of economic development, the trade cycles or capitalism,[13] which explains the phenomena in question. Schumpeter's work in this field must be regarded as really only a very ramified analysis and description of the phenomenon of innovation and its consequences.

Nevertheless, it is difficult to overestimate the stimulating effect exerted on economic theory by Schumpeter's view of the significance of innovations, and the concept of bursts of innovations has in fact stimulated both empirical and theoretical research.

Schumpeter states expressly that innovations are not the results of inventions, but instead the capitalistic mode of production, characterized by the introduction of innovations, creates a suitable climate for the invention of new techniques. As his examples show, these innovations do not only consist in the application of inventions. The inventive process as such is thus entirely secondary to the application of innovations. In Schumpeter's view there is always a large reservoir of possible technical processes available, which only need creative entrepreneurs to put them into practice.

This casts doubt on some authors' view that Schumpeter's notions imply that innovations depend on the development of technology.[14] This view is contradicted not only by Schumpeter's own refutation, but also by the fact that he includes, for example, the opening-up of new markets among innovations, and particularly by the fact that—according to him—the actual application of new combinations depends on entrepreneurs with certain psychological characteristics and on the profitability of new ventures. When one uses Schumpeter's interpretation of innovations, it is particularly unjustified to say that he was concerned with the social and economic aspects of technical change in the sense of the effect of new technical possibilities on social and economic development.

The main reason why it is believed that Schumpeter pays so much attention to technical development is the generally accepted view that Schumpeter makes an excessively sharp distinction between innovations and inventions. Solo points out that 'invention and innovation are normal business activities,[15] and it can be seen from the account of technical development up to 1900 that inventions and their applications are closely intertwined, even though the large-scale application may come long after the invention. High costs of innovation may urge one to improve the original invention; and the application of an invention can lead to new inventions. The regular alternation between invention influencing innovation and innovation influencing invention suggests that development occurs not so much in bursts as in a continuous stream of new technical possibilities and applications. Strassmann concludes from a detailed study of American industry in the nineteenth century that the main objection to

Schumpeter's theories is that 'they do not adequately explore the process of technological change as a series of complementary, mutually reinforcing developments'.[16] Since inventions also partly depend on the funds available for research and development the demarcation between inventions and applications becomes blurred, as does the distinction between the technical and other aspects of economic development. Taking all this into account, one can understand why Schumpeter's name is often mentioned in connection with the development of technology.

In addition, Schumpeter himself treats innovation as the creation of a new production function, which is often regarded nowadays as a characteristic feature of technical development. Schumpeter overlooks the fact that a displacement of the production function cannot be automatically connected with some of the phenomena he called innovations, for example the opening-up of new markets. Furthermore, a critical assessment of his ideas is not made any easier by his different interpretations of the production function. According to his subjectivistic interpretation, the production function expresses the technical possibilities that lie within the scope of the entrepreneur. This implies that a new production function is created whenever the entrepreneur increases his technical knowledge. The practical application of techniques known to him does not shift the production function, so that according to this subjectivistic interpretation an innovation does not give rise to a new production function. Schumpeter's second interpretation is that the production function fully reflects the state of technology. Here again, the application of an available technique does not lead to a new production function. In his terminology, only an invention—and not an innovation—will give rise to a new production function. This is compatible only with his third interpretation of the production function, according to which the production function reflects the combination of production and the factors of production in practice. Then the introduction of a new combination does mean that the production function is displaced.

The objection to this last version is that the production function no longer gives a survey of the technical alternatives, but reflects instead the entrepreneur's decision, based on technical and economic considerations. Furthermore, since in the case of unchanged technology a change in relative factor prices also causes a shift in the production function, the constant change in the production function indicates an economic development rather than a technical one.

If Schumpeter's second interpretation is accepted, an innovation does not cause a shift in the production function, because it does not involve a change in the technology. Since this second interpretation is the dominant one in economic thought, it is not surprising that the view that Schumpeter dealt in depth with technical development is so widespread. If one adopts the notion that the practical application of a known technique leads to greater technical knowledge, then innovation can give rise to a new production function, again according to Schumpeter's second interpretation. In that case, however, his distinction between inventions and innovations is lost.

Schumpeter, therefore, did not really analyze the development of technical possibilities and their effect on economic life, despite his often fascinating account of what he calls the dynamics of capitalism. The central features in his theory are the emergence of entrepreneurs with certain psychological characteristics[17] and the effects of their innovations, rather than the more objective aspects of the technology of production. In this sense he belongs to the Austrian school.

Although both Marx and Schumpeter believed that capitalism would change into socialism, they produced very different reasons for this view. According to Marx, the introduction of new labour-saving techniques leads to a redundant labour force, because the accumulation of capital is not sufficient to ensure re-employment. Innovations, determined by the development of technology, occur independently of the capitalist's or entrepreneur's personality and eventually bring about the collapse of the system. In Schumpeter's work, on the other hand, it is only the entrepreneur's personality that explains the new combinations and constitutes the essence of capitalism. As the ability to achieve new combinations becomes commonplace and innovations become a routine matter, the unique characteristics of the entrepreneur disappear. In Taylor's words, the mainspring of the dynamics of capitalism, according to Schumpeter, is the 'creative spirit of the great entrepreneurs; and capital growth is a response . . . to the growth of demand or investment outlets first created by the entrepreneurs, not the other way round'.[18] In Marx's theory, on the other hand, the entrepreneur is 'simply an active capitalist who is busy in creating and augmenting surplus value in simple as well as expanded reproduction'.[19] Schumpeter's subjectivistic picture of change in the economic system therefore contrasts with Marx's objectivistic view.

6.4 CONCLUSIONS

The conclusion that Schumpeter was only incidentally concerned with technical development does not detract from the special position accorded to him in this field. His concern with the changes in economic life showed up the limited value of static equilibrium theory, and his vision is still a stimulus to economic theory and research. We shall now mention briefly three problems which will arise in later chapters and which have been influenced, both in their formulation and in their treatment, by Schumpeter's highly individualistic approach.

Schumpeter's sharp distinction between innovations and inventions has triggered off several reactions, and both theoretical analysis and empirical research have been directed to the question whether innovations are really as independent of inventions as Schumpeter supposed. A critical evaluation of Schumpeter's theory has led to a discussion of the nature of technical development, the length of the period between an invention and its application, and the diffusion of new techniques. Mansfield, who has made some important contributions in this field, remarks that, quite apart from the question of whether an invention and its application should be so sharply separated, 'innovation is a key stage in the process leading to the full evaluation and utilization of an invention'.[20] Nordhaus has developed a mathematical model in which the stream of inventions is explained by economic factors and thus disagrees in this respect with Schumpeter, who emphasizes non-economic factors.[21]

The second problem concerns the interpretation of the production function. Salter's study of productivity and technical development, which has already become a classic, contains a penetrating discussion of the production function, which was partly inspired by Schumpeter's thoughts on the subject in his *History of Economic Analysis*.[22] Hicks' distinction between production functions that express the relationship between equilibrium values of production and the factors of production, and the currently accepted form of the production function, also relies on an interpretation which we have encountered in Schumpeter's work.[23] The connection between technical development and the production function will be discussed in the next chapter in more detail.

Schumpeter's thoughts on the effect of monopolistic and oligopolistic markets on technical change have given a great impetus to the study of perhaps the most important question in our field, namely the relationships between technical change and the market structure. For example, in his well-known study on oligopoly and technical development, Sylos-Labini repeatedly refers to Schumpeter.[24]

Although modern growth theory, in which technical development occupies an important position, does not adequately represent the 'Schumpeterian System,'[25] Schumpter's work has had a positive and stimulating influence on current microeconomic growth and production theory,[26] by virtue of his hypotheses which are mainly based on historical observations—despite the fact that his work often exhibits a conspicuously tautological character. The next chapter, about technical development since 1900, will help us establish whether Schumpeter's generalizations are valid for the twentieth century.

7 Technical development since 1900

There is every reason to believe that we should be able to renew our human bodies in the same manner as we renew a defect in a boiler.

Henry Ford[1]

7.1 INTRODUCTION

The moon landing in the summer of 1969 is a dramatic example of technical advances made in this century. This highly advanced and spectacular application of technical knowledge would not have been possible without a high level of scientific, medical and mathematical knowledge. Technical experts from various fields worked together in teams to build the necessary equipment and to guide the astronauts throughout their journey. A single defect in one of the millions of components could have meant failure for the whole Apollo project. Several computers were needed, for example to calculate the correct trajectory so that reliable mid-course corrections could be made. Television transmissions from space added a mysterious dimension to this event. Besides scientific motives, military considerations also play a part in reconnoitring space. The fact that the earth's and the moon's gravity could be overcome implies that very powerful sources of energy are now available. The landing on the moon and the return of the astronauts to earth are only some of the problems whose solution could be envisaged in advance.

We shall deal here with only some of the modern technical achievements because comprehensive treatment would require a separate study.[2] Section 7.2 deals with technical development in general terms. We then discuss the connection between technology and war, since the history of the twentieth century has been dominated by two world wars and the rate and type of technical development are definitely influenced by military concerns. The focus then turns to the field of electronics, not least because of the great and growing importance of the computer. This is followed by some remarks on automation and the space programme.

An investigation of these results of modern technical development creates the right background for some predominantly theoretical considerations and a closer analysis of Schumpeter's work in the following chapters. Furthermore, the expression of technical development by a production function raises numerous questions which can be more successfully tackled with concrete knowledge of some representative examples of technical development.

7.2 A GENERAL OUTLINE

Several economists have pointed out that technical development changes not only the volume but also the structure of production.[3] The contribution of the three sectors of agriculture, industry and services to overall production in developed countries is decreasing in the case of agriculture, expanding fast in the case of industry, and expanding gradually in the case of services, with technical development playing different roles in each sector.

Due to mechanization and chemical science the productivity of labour has steadily increased in the agricultural sector in this century. The decrease in the share of agricultural produce in total production can be explained by the small income elasticity of the demand for agricultural products.[4] Mechanization and chemical technology have led to a spectacular increase in production and the productivity of labour in industry, and a considerable part of total demand has so far been directed at industrial goods. The introduction of the computer and data processing systems has made possible the automatic control of industrial production processes. According to Crossman's estimates, about one per cent of all mental processes were automated in 1966 and the automation of all mental processes will take a hundred and fifty years.[5] These estimates, based on the careful examination of a great number of processes, refer not only to industry but also to services, for here too the use of the computer is gaining ground, for example in banking and insurance. A substantial increase in productivity can be expected only when people fully accept and use these new developments. But already the composition of GNP is slowly shifting from industry to services.

The disappearance of old sources of energy and the tapping of new ones are changing not only the composition of production, but also the nature of the production process. While the emphasis has so far been on the use of physical energy, it now seems that the acquisition, storage and processing of information—i.e. mental energy—are gaining ground. We have seen in Chapter 4 that in the course of hundreds of years man gradually came to supplement his muscle power with the use of tools and devices. The great impetus came in the eighteenth and the nineteenth centuries when people began to utilize fossil fuels like coal, oil and natural gas as sources of energy. With the harnessing of atomic energy, forces hitherto undreamt of are being tapped. Various devices are now being introduced to replace mental processes. Although there were instances of automatic control before, for example Watt's governor, it is only now with the development of electronics that the automation of production can make significant strides.

7.3 WAR AND PEACE

The first half of this century witnessed two world wars in which millions of people died, while the second half has been dominated by an arms race. The ceaseless renewal of military apparatus eats up a large slice of the world's resources. The potential annihilation of mankind now possible explains why the terms 'technical development' and 'technical change' are more apt than 'technical progress'. (Although the fact that there has not yet been a third world war may well be due to the existence of the hydrogen bomb.)

Some people believe that the economy in the West is kept afloat by military expenditure and the military contracts given to industry. While I do not accept this extreme view, it is undeniable that military expenditure introduces a more or less stable demand factor into economic life, and also that this military expenditure influences both the rate and the type of technical development.

It is well known that war moulds technology. Emergency situations produce inventions, and productive forces are turned towards developing new weapons and military communication systems. Every war intensifies technical development. Under the pressure of circumstances, new technical possibilities are discovered in a relatively short time, and then put, sometimes almost immediately, into large-scale production. After the war the technical advances made can often be put to peaceful use, and the period after the Second World War illustrates this particularly well. The post-war development of aviation and space exploration, and the applications of nuclear energy and electronics illustrate the rapid and large-scale civil use of military technologies. In such a wealth of new discoveries, a distinction can be made between imitation and application by analogy.[6] In the first case, no special adaptation is necessary for the diffusion of the technical invention, while in the second case the invention must be adapted to an originally unforeseen purpose before it can be widely applied.

The complexities of the diffusion of technology are illustrated by the case of the computer which was partly developed and perfected for defence and space programmes—computers being important for controlling rockets and ballistic missiles. Electronics, considerably influenced by the arms race and space exploration, has found numerous applications in the consumer goods sector.

7.4 ELECTRONICS AND THE COMPUTER

We have already mentioned that an electric current can generate a magnetic field. The opposite case was demonstrated by Sir Michael Faraday (1791–1867), who generated an induction current in a closed conductor by varying the magnetic field enclosed by the conductor. This discovery of electromagnetic induction was largely instrumental in making possible large-scale applications of electricity.

After Maxwell (1831–1879) produced, on a theoretical basis, equations for calculating the velocity of propagation of electromagnetic waves, Hertz (1857–1894) succeeded in demonstrating these waves experimentally. While the range of Hertz's observation was small, Marconi (1874–1937) developed some methods of transmitting electromagnetic waves over much greater distances, using diodes and triodes in the receiver. This opened up the way for radio transmission and telegraph and telephone systems without wires and cables. Television and radar similarly rely on the electromagnetic wave theory.

These developments in the field of electronics are a good illustration of the technical advances made in this century. In many cases discoveries based on theoretical scientific principles have had spectacular large-scale practical applications. Enormous sums of money have been allocated to the development of promising projects. One of the results of this is the transistor, which replaced traditional radio valves such as diodes and triodes after the Second World War.

Although mechanical computers have been developed, the rise of the computer has been closely connected with electronics. Mechanical switching components are slow, electromechanical ones fairly fast, and electronic ones very fast. The first electronic computer was switched on in 1946 and was, for example, used for ballistic calculations. This computer contained 18,000 electronic and 1,500 electrical relays, which shows how heavily computers of this type rely on electronic principles. Most computers operate on the binary principle, with on and off modes, and can be regarded as very fast, automated calculating machines with a large memory. Automation in this case means that computational and administrative steps are carried out automatically one after the other.

For these operations, the computer is fitted with a memory or store. A compact store with a rapid access is nowadays usually built up of small magnetic loops which exhibit a binary behaviour by virtue of the North and South poles. The size of computers is greatly reduced by replacing electronic valves and transistors by integrated circuits. For the long-term storage of data, other types of stores are used, such as magnetic tapes, drum stores and, increasingly, disk stores.

Computers cannot think for themselves and are entirely dependent on the instructions they receive. They must be programmed in a language they understand, i.e. in a computer language. The computer only knows what the programmer tells it, but its possible applications are practically unlimited. Space exploration is unimaginable without the computer; computers control numerous industrial production processes; many administrative processes are carried out by computers in an automatic manner; many mathematical calculations can be performed very quickly by computers; and there are now numerous uses of a computer's memory, for example in the field of documentation.

The introduction of computers into business life,[7] education and government activities—in fact, into the general life of society—is bringing about a profound change in the organization of production, the position of employees, the incentive for the accumulation of capital, the increase of productivity and the creation of leisure time. The essential difference between the computer and other machines is thus its very wide variety of applications, due to the flexibility of the programs that can be written for it.

7.5 AUTOMATION

There have been many different descriptions of automation. Emphasis is sometimes placed on sophisticated technical apparatus or the special role of data processing, while in other cases the consequences of automation are included in the definition.[8] Long before the introduction of the computer, parts of industrial processes were automated;[9] machines performed automatically a series of operations and—at a more advanced stage of development—also responded automatically to the information fed into them. But this was only partial automation. In a fully automated plant, production, handling of materials on the premises, and supervisory operations are all carried out in a self-regulating manner. Though this can in principle be imagined without the computer, in fact only computerized data processing has made it possible to integrate these three operations, for only it can coordinate the information to and from these operations.

According to this definition, automation is different from mechanization, where the replacement of human muscle power by machines is the central issue. However, the transition between the various stages is not sharp. Bright has investigated the various levels of mechanization and automation in his famous work *Automation and Management*, where he distinguishes between seventeen levels, encompassing simple manual work, the use of simple mechanical devices, the use of machines that respond to the materials fed into them and various types of self-regulating machines. Mechanization gradually changes into automation, accompanied by a profound change in man's position in the production process, at least in retrospect. Direct contact between man and machine is replaced by a control system which operates independently some distance away, as for example in the petroleum industry. This sort of automation is the use of 'advanced technical means or equipment in fully or mostly self-regulating processes, which leads to a great saving of labour and a considerable increase in the productivity of labour.[10]

However, though automation does cause a profound change in the position of the employees, the description of automation above is not completely satisfactory, since one could equally emphasize other important consequences of automation. These are, for example, the control of the quality of the production process and the end-product, and the control of the flow of information. Automation can lead to new products and services, such as withdrawing money from a bank outside working hours and, in due course, the use of the Giro system by telephone. The increase in the productivity of labour is, therefore, not the only important aspect of automation. Other aspects include the very large investments involved in automated systems. However, this objection does not detract from the above definition if we realize that the aim was the formation of labour-market policy. The economic, social and psychological effects exerted by automation on employees are often so dominant that one should not be surprised that they were singled out in the above definition of automation given by the Dutch Socio-Economic Council.

Generally, automation alters the demand for various types of labour without causing large-scale redundancies; it reduces the demand for unskilled labour and raises the demand for skilled labour both in absolute and relative terms. The demand for specialists in automation and data processing rises steeply, as does the demand for managers. Automation thus rearranges the staff composition, and there are often objections.[11] Retraining should not therefore be aimed only at imparting new technical skills to people, but also at instilling flexibility in them. The latter applies particularly in the case of older employees, whose livelihood often seems to be endangered by automation; if there is no flexibility in such cases, the temptation may be to pension people off before they reach retirement age. Furthermore, the automation of administrative processes, in particular, can change the pattern of demand for various types of female labour. For example, the demand for part-time workers in the Dutch Post Office Giro Service greatly decreased, while the demand for punch-card operators has increased.[12]

Automation—including that of administrative processes— will undoubtedly continue both in industry and in the public sector. Rapidly rising wage bills often act as an important incentive to accelerated mechanization and automation, but the process is restricted by limited human flexibility and an insufficient supply of skilled labour. The serious consequences of small technical mistakes or problems makes an automated system a difficult and complicated business to install. Besides, one cannot ignore the widespread fear of robots and the violation of privacy associated with certain forms of automation, together with the fear of redundancy. Automation is associated with a great deal of uncertainty, which can easily lead to unrest, labour conflicts and other social disturbances.

7.6 SPACE EXPLORATION

The detailed description Jules Verne gave in 1865 of a journey to the moon by three astronauts, Barbicane, Ardan and Nicholl, is remarkably close to the actual journeys to the moon that have taken place. This French writer had an unshakable faith in technical progress. The merging of fact and imagination, which is characteristic of his work, can be seen from the following passage, in which he discusses the geological importance of the investigations of the moon:

The expedition of Barbicane and his friends round the moon enabled them to correct the many current theories regarding the satellite. These savants had observed *de visu* and in unique circumstances. Now it was clear which systems of thought should be rejected and which retained with regard to its formation, its origin, its habitability. Its past, present, and future had given up their last secrets. Who could advance objections against conscientious observers who at less than twenty-four miles distance had studied the mountain of Tycho, the strangest system of lunar orography? How answer those savants whose glance had penetrated the abyss of Pluto's crater? How contradict those bold ones whom the chances of their enterprise had borne over that invisible face of the disc which no human eye had before seen? It was now their turn to impose some bounds on that Selenographic science which had reconstructed the lunar world as Cuvier did the skeleton of a fossil, and to say, 'The moon was this, a habitable world, inhabited before the earth! The moon is that, a world uninhabitable, and now uninhabited.'[13]

When Jules Verne was writing man had already gone up in a balloon but not in an aeroplane. The first artificial satellite was launched by the Soviet Union in October 1957, and after that the Americans rapidly made up for their initial backwardness in this field. Accurate photographs of the moon surface were taken from a short distance in the summer of 1964, and some unmanned satellites performed soft landing exercises on the moon in 1966.

A space-craft with astronauts on board was in orbit round the moon at Christmas 1968, making valuable observations, for example of the other side of the moon, which was then directly seen by man for the first time. The first landing on the moon by man took place in the summer of 1969, about a hundred years after Jules Verne's fantasy. This was followed by other missions, in which the astronauts each time stayed longer on the moon and made increasingly thorough investigations. The remarkable thing about this development is that so many difficult technical problems were solved in such a short time, problems such as escaping from the earth's gravitational field, landing on the moon, take-off from the moon's surface, re-entry into the earth's atmosphere and recovery from space. The great speed with which these problems were solved, mainly by the Americans, and the precision of operations which was achieved indicate that, at least in this field, technical development is the outcome of a systematic programme in which personnel and material are directed towards a specific goal.

Space research is not confined to the moon, and journeys are planned to Mars, Venus and other planets. Satellites provide useful information for weather forecasts (particularly long-term ones) and are useful for worldwide telecommunication purposes. The incentive of military applications has significantly encouraged space research, and orbiting satellites have opened up virtually unlimited possibilities of espionage.

In the field of space exploration ends precede means. Space projects—including the research work necessary for them—need enormous sums of money, which could of course be used for other purposes. Technical development in this field can be stimulated by attracting experts and placing large sums of money at their disposal, and it is equally possible to slow down the rate of development by cutting down on expenditure and personnel.

7.7 INVENTIONS AND INNOVATIONS

Innovations are sometimes the precursors of inventions, and sometimes their outcome. The application of an invention sometimes follows directly from a discovery (which in this century is based mostly on new scientific knowledge), while in other cases there is a considerable time-lag between the invention and its application. Even when a clear idea exists of the possible applications, these may be hampered by commercial, social or technical factors.

There may also be a clear idea of an innovation but the invention needed for realizing it has not yet been made; the history of space exploration contains several examples of this. The envisaged innovation is here a precursor of the invention, and when the invention thus induced is put into practice, new and unexpected discoveries may result from it. The reduction of the size of computers by the use of integrated circuits illustrates the saying that necessity is the mother of invention.[14]

The history of technical development in this century points to a continuous stream of inventions and innovations which may be intertwined without always being necessarily connected causally or sequentially in time. The sudden revolutionary spark of the first insight is often outstripped by the gradual process of accumulating, shaping, perfecting and applying new technical possibilities. Inventions always increase the body of technical knowledge, but innovations can also do this; it is therefore one-sided to restrict the description of technical development to inventions only.

Our brief sketch of some of the technical achievements in this century shows that it is not possible to make any categorical statements about the relationship between inventions and innovations, between theory and application, and between science and technology. The variety of interactions involved in technical development reflects human inventiveness and intellect, besides the variety and complexity of human affairs and the playfulness of the human spirit.

Detailed and careful investigations by Jewkes, Sawers and Stillerman have also revealed that the stream of technical achievements is fed by very different sources. These authors have studied numerous inventions and the lives of many inventors and have drawn some well-founded conclusions. For example, they found that in the nineteenth century there was more contact between scientists and inventors—and more teamwork—than is usually assumed, in an economy where there was still virtually no 'large-scale systematic research in a field where results are *a priori* likely'.[15] Most inventions in the nineteenth century were not foreseen[16] and even the application of new scientific discoveries—often by the discoverers themselves— came as a surprise. These authors' account of invention and innovation made in the period 1958–68 suggests that (a) independent inventors still play an important role alongside institutionalized research, (b) inventions are made by enterprises of various sizes, (c) most inventions consist of relatively small improvements and (d) economic factors constitute the most decisive determinant of the stream of inventions, although other considerations— including chance—also affect it.[17]

The following conclusions can be drawn from empirical investigations by Enos,[18] Lynn[19] and Mansfield. In the areas these authors studied, the most important innovations follow inventions after an average time-lag of between ten and fifteen years, the standard deviation from this being a period of five years in the petroleum industry and sixteen years in various other fields. The time-lag is shortest with mechanical innovations, and longest with electronic ones; furthermore, it is shorter for consumer goods than for industrial ones. Inventions financed by the public sector are applied after a shorter interval than those financed by the private sector. On the whole, the average time-lag seems to be decreasing. The diffusion of an innovation to other enterprises in the same sector is generally a slow process, not infrequently taking twenty years. The rate of imitation also varies greatly, and the finding that innovations now diffuse somewhat faster than before may be purely fortuitous. Extensive diffusion of an innovation sometimes takes scores of years, while in other cases the company which first introduced the innovation is quickly followed by others.[20] However, it appears from a detailed study[21] that the diffusion of technical knowledge is an important part of the development of technology, and that it mainly stems from the movement of creative technical people from one job to another. Some interesting distinctions have been made in recent literature on the transfer of technology. Thus, Mansfield[22] has stressed the difference between the vertical and horizontal transfer of technology. The first goes from basic research to applied research, then to development work and finally to production (and in the opposite direction), while in the second case applied technical knowledge is transferred from one place or enterprise to another. It is also possible to distinguish between general technical knowledge (common to an industry), system-specific technical knowledge (specific to the manufacture of a certain product) and firm-specific technical knowledge (specific to a certain firm[23]). Furthermore, it seems important to distinguish between various stages in the diffusion and application of technical knowledge.

It can be concluded from all this that Schumpeter's distinction between invention and innovation has been very fruitful. It has enriched economic theory and stimulated empirical research. However, both theoretical and empirical observations indicate that the relationship between inventions and innovations is more varied and complicated than Schumpeter thought. In particular, technical achievements since 1900 show how untenable is the notion that innovations are introduced in bursts independently of the corresponding inventions and immediately find widespread application.[24] We are therefore inclined to see in Schumpeter's argument a formalized expression of the process of technical innovation that occurred in the nineteenth century under the influence of some spectacular and well-known inventions. In general, it is more realistic to view the alternation of invention and innovation as a continuous stream, although this is not meant to imply a causal relationship.

7.8 CONCLUSIONS

The advances of technology have left many people in this century dumbfounded. The disappearance of the horse-drawn omnibus and gas lighting, and the coming of the car, aeroplane and electrical lighting are events to which people have had to acclimatize themselves, not always without difficulties. Yet people have adapted remarkably quickly to visual and audial communication systems which have made simultaneous worldwide reporting possible within a very short time. It may be that visionary writers like Jules Verne have prepared us mentally for space travel more than we appreciate. While Jules Verne's fantasies had to wait a hundred years before they came true, Henry Ford's prediction about the development of medical techniques, quoted at the head of this Chapter, has become reality within half a century.

The increase in technical knowledge can be embodied in capital goods, but this is not necessarily the case. One and the same invention, for example television, can act as the centre of a wide network of applications without each new application requiring the creation of new capital goods. Technical development can be triggered off by intensive institutionalized research work, but it may also come from individual and independent inventors. Advances in the case of the computer, automation and space exploration illustrate the dependence of technology on science, but important improvements can still come about quite spontaneously.[25]

A certain state of technology, characterized by a set of technical possibilities, can in principle be described with the aid of a production function in such a way that each of the aspects of technical development mentioned above is taken into account separately. However, now that we have in general rejected the notion that inventions and innovations occur in bursts with innovations always following inventions, and have instead adopted the view that technical development is a continuous process in which the technical possibilities increase, we are faced with the question of what causal connection exists between the events in this sequence. This becomes the question whether a historical series of production functions, each of which gives a description of the state of technology at a given moment in time, can be taken as a concatenation that also expresses the causal connections between the events of technical development. This question is central to the theory of production.

8 Microeconomic production theory

As we succeed in recognizing and incorporating one aspect of the real world in our models, our failure to incorporate other aspects becomes more apparent.

T. C. Koopmans[1]

8.1 INTRODUCTION

We have seen that there is no place for an exact interpretation of the production function in Marshall's work,[2] and—connected with this—the law of increasing and diminishing returns within a firm is not clearly distinguished either from the phenomenon of external economies of scale or from a situation where all the factors of production in the firm are changed. It is not without reason, therefore, that Samuelson criticizes this aspect of Marshall's work, which has produced what is known as the 'cost controversy'—a topic that exercised the minds of economists in the period 1920–40.[3] This mainly concerned the shape of the cost curves in the short and long run, and, thus, the shape of the individual and market supply curves in the long run.[4] Another important question was whether an equilibrium position could be formulated in the case of perfect competition when the cost curves had different shapes. This came to a head in the case of decreasing average costs, which—combined with a horizontal individual demand curve—would lead in theory to unlimited expansion of production and sales. Sraffa pointed out in 1926 that the possibilities of selling products are limited, which implies a downward sloping individual demand curve. This led to a revolution in price theory in the 1930s, as a result of which attention shifted from perfect competition to monopolistic types of competition.[5]

While the economists involved in the cost controversy used mathematical tools only to a limited extent or not at all, other economists, who followed for example in Wicksell's footsteps, proceeded to express the relationship between the cost function and the production function in a mathematical form. While

Stackelberg[6] concentrated mainly on the analysis of the cost function. Schneider presented a lucid survey of the exact relationships between cost and production functions under various types of competition.[7] Carlson[8] did the same a few years later, but while Schneider assumed that production is a timeless process, Carlson assumed that the firm acquires the means of production at the beginning of the period under consideration and sells the products at the end of this period.[9] Using Schneider's and Carlson's approach, it is easy to see that under perfect competition the individual equilibrium is indeterminate if one assumes constant returns to scale.

Besides the predominantly verbal and limited mathematical approach to the microeconomic aspects of production, mention must be made of a more empirical macroeconomic approach, described in a pioneering article by Cobb and Douglas, which appeared in 1928.[10] These authors explained the growth in production in the period 1899–1922 by the increase in population and in the stock of capital goods; they expressed the increase in the fixed capital C, in the population L and in the production P by a series of index numbers, and proposed a relationship of the type

$$P = b.L^k.C^{1-k}$$

where, on the basis of econometric investigations, $b = 1.04$ and $k = 3/4$, the correlation coefficient being about 0.95. Cobb and Douglas's article ends with a discussion of the significance of this macroeconomic production theory for the distribution of the national income, their conclusions being that the processes of distribution follow in large measure the processes of production if sufficient time is allowed. In this respect, the contribution of Cobb and Douglas is a further development of the attempt, started by Wicksteed, to combine the theory of production with the theory of distribution via the theory of marginal productivity, and it also retains the macroeconomic character of this approach. It is interesting to note that technical change in the period under consideration is not taken into account. Tinbergen was among the first to improve the analysis from this point of view,[11] and his work was followed, after the Second World War, by a series of sophisticated attempts to incorporate technical change in a dynamic macroeconomic production function.

However, before dealing with these empirical studies, we shall examine the theory of production more closely and discuss its microeconomic aspects in this chapter and its macroeconomic aspects in the next. This will enable us to tackle the complicated problem of the existence of a macroeconomic production function after studying the various interpretations and formulations of the microeconomic production function.

8.2 INTERPRETATION OF THE PRODUCTION FUNCTION

Empirical production functions, which link the actual, recorded output with the actual factor inputs, should be sharply distinguished from theoretical production functions, which describe the connection between production and the factors of production *ex ante*. In the latter the production function tells us what technical possibilities are available to the firm. In a simple model, a combination of the factors of production is then chosen that reflects the objectives of the enterprise and the relative prices, and there is of course a corresponding scale of production. In principle, such equilibrium combinations are reflected in the empirical data. In empirical investigations, however, by studying the behaviour of generalized, *ex-post* combinations of production and its factors over time, one examines a whole set of relationships, of which only one component is the theoretical production function.

We can divide theoretical production functions into subjectivistic and objectivistic types. The subjectivistic variant comprises the technical possibilities known to the firm, so that the production function expresses the technical knowledge that the firm possesses. The production function then does not reflect the general state of technology, since this may be more advanced. In the objectivistic function, on the other hand, the overall state of the technology is reflected. If a complete and instantaneous diffusion of technical knowledge is assumed, there is no difference between the two interpretations of the production function, but this assumption is generally not possible in practice, partly because the diffusion of technical knowledge is inevitably slow and partly because it is deliberately opposed by enterprises which incurred financial costs to acquire the new technical knowledge. By applying for a patent, the firm relinquishes secrecy, but this does not promote the diffusion of technical knowledge from the point of view of the society. The effects of this on social welfare will be discussed in Chapter 12.

The question therefore arises whether the term 'technical development' should be used to mean an increase in technical knowledge in the subjectivistic sense, giving rise to a new subjective production function, rather than in the more usual objectivistic sense. It is useful to start from the view that the diffusion of *existing* technical knowledge is different from the acquisition of *new* technical knowledge. If we include the diffusion of technical knowledge in the term 'technical development', an invention made within a firm that increases technical knowledge must be classed with the acquisition of technical knowledge from outside and, as it is important to distinguish between these two processes, we must keep them separate in our terminology. Therefore, technical development in the narrower sense is not equated with the diffusion of technical knowledge, since it is only from the standpoint of the firm under consideration that a new production function arises here. We can speak of technical development in the narrower sense only if the technical knowledge of the whole society increases in absolute terms. A change in the state of technology can then be regarded as the appearance of a new objectivistic production function.

It is useful to make a sharper distinction than is customary between an increase in the range of technical possibilities and an increase in technical knowledge; this is what underlies the distinction made in Section 1.2 between technical development in the narrow sense and that in the broad sense. The latter term is used here to mean the diffusion of technical knowledge, since the firm acquires new technical possibilities by obtaining already existing technical knowledge from elsewhere, so that the technical possibilities increase also in the objective sense. This view is shared by Hicks,[12] who is also inclined to equate a diffusion of knowledge with technical development. The production function reflects the technical possibilities, which do not depend on technical knowledge and its diffusion alone. This will be examined more closely in Section 8.7, together with the question of whether technical development can be equated with a shift in this production function.

The production function portrays all the efficient production techniques existing at a given moment in time.[13] These techniques form a mathematical set of combinations of the means of production that ensure the maximum output of a certain product.[14] There is a single maximum output belonging to a certain combination of the means of production. The production function is the expression which translates the quantities of the means of production into the quantity of output produced.

Production functions belong to one of two groups: in the first group it is assumed that the means of production are completely divisible and fully interchangeable, while in the second group one or both of these assumptions are not possible.[15] Of this second type, only one, known as Leontief's production function, will be examined here, namely that in which the means of production are visible but not interchangeable and in which the technical coefficients are fixed. The first group comprises both the classical production functions, which differentiate between a period of increasing returns and a period of diminishing returns, and the neo-classical production functions which express only diminishing returns. Both increasing and diminishing returns assume a situation where a variable factor of production is combined with a set of production factors which is kept constant. The other well-known distinction—between increasing, constant and diminishing returns to scale—is connected with the variation of all the factors of production.

8.3 FORMULATION OF THE PRODUCTION FUNCTION

The Cobb-Douglas production function, the empirical variant of which was mentioned before, is a much-quoted neo-classical production function. If for the sake of simplicity only two factors of production, capital and labour, are considered and their quantities are denoted by K and L, and if the amount of production or output is denoted by Q, then this production function has the following form:

$$Q_t = a.K_t^{\alpha}.L_t^{\beta} \qquad a > 0,\, 0 < \alpha < 1,\, 0 < \beta < 1$$

This production function describes the technical possibilities available to the entrepreneur at time t, and therefore it embodies the state of technology at time t.

It can be shown that the marginal products of capital and labour are positive, and that for feasible values of α and β, they decrease. The exponents α and β are the corresponding elasticities of production. The returns to scale increase, remain constant or decrease depending on whether the value of $\alpha + \beta$ is greater than, equal to or smaller than 1. The sum of the two elasticities of production is equal to the elasticity of scale, this relationship being known as the Wicksell-Johnson theorem. Finally, it can be shown that the elasticity of substitution is always equal to -1, provided that the remunerations are assumed to be directly proportional to the marginal products of the factors of production.[16]

Technical development can manifest itself in different ways in the Cobb-Douglas production function. Firstly, it can be mirrored by an increase in the coefficient a, which means that the maximum amount Q is higher, though the combination of the factors of production has remained the same.[17] Such a technical development is neutral in Hicks' sense. If the other coefficients remain the same, there is a shift in the graph of the production function. Technical development may also change the elasticity of scale $(\alpha + \beta)$, but a difficulty here is that this can also be brought about by a change in the size of the firm. It would therefore be advantageous to have production functions in which changes in scale can be distinguished from technical development. Furthermore, technical development can also change the elasticity of production; production methods can become variously capital-intensive for the whole set of combinations of capital and labour, so that one is faced with a Hicksian, non-neutral technical change. Technical development cannot manifest itself in a change in the elasticity of substitution, since this has a constant value of -1. In short, technical development can change one or more of the coefficients in the Cobb-Douglas production function, but it is also conceivable that it leads to a completely new production function. It should also be noted that a change in the coefficients of the Cobb-Douglas production function is not always due to technical development.

A production function that gives more scope to technical development is the 'constant-elasticity-of-substitution' (or *CES*) production function, of which the Cobb-Douglas one is a special case. An essential feature of this production function—as its name suggests—is that the elasticity of substitution has a constant value, which is not given *a priori*.[18] Using the same symbols as for the Cobb-Douglas production function, we can write the CES function in the following form:

$$Q_t = a[bK^{-\alpha} + (1-b).L^{-\alpha}]^{-\beta/\alpha} \quad a > 0, \ 0 < b < 1, \ \alpha > -1, \ \beta > 0$$

It can be shown that the marginal products of capital and labour are positive and that for feasible values of β they decrease. The exponent β reflects increasing, decreasing or constant returns to scale. The elasticity of substitution σ is equal to $1/(1+\alpha)$. It can be seen furthermore that no finite limit is put on production if labour is combined with a constant amount of capital in the case where $\sigma > 1$, but if $\sigma < 1$ production reaches a finite maximum.[19] The explanation for this is that when it is difficult to substitute one factor of production for the other, the factor that is kept constant hinders the growth of production; when such a substitution is easy the constancy of one of the factors can be offset by substitution.

Let us now examine the influence of technical development in the case of the CES production function. If technical development raises the value of the coefficient a, we are again faced with Hicks' neutral technical development, and the production function is shifted in a simple manner. Such a situation arises also when the elasticity of scale β is changed by technical development. It should be noted that in this case, too, the value of β can be changed by variations in the size of the firm. Hicks' non-neutral technical change can alter b or α, or both. If the exponent α is affected, the elasticity of substitution σ changes. As in the case of the Cobb-Douglas production function, therefore, technical development can be manifested in a change in the values of the coefficients of the CES production function, but again it can just as easily lead to an entirely new production function.

If the assumption, made so far, that the factors of production are mutually substitutable is replaced by the assumption that they can be combined only in one way to obtain a certain output efficiently, then the technical possibilities are described by Leontief's production function.[20] In the case of constant returns to scale, the output is directly proportional to the critical factor of production that is acting as a bottle-neck; the marginal product of this factor of production is positive and constant, while the marginal product of the other factor is zero. In this case, technical development changes the values of the fixed technical coefficients in Leontief's production function.[21]

The neo-classical production functions in particular have been generalized in the last few years from various points of view. Thus production functions have been developed with a variable elasticity of substitution (VES) and an elasticity of scale that depends on the level of output (for example the VES function).

8.4 INTERTEMPORAL PRODUCTION FUNCTIONS

The incorporation of the concept of the production period into the production function leads to an intertemporal production function. Wicksell's description of Böhm-Bawerk's theory in the form of a relationship between the per-capita production q and the production period t (cf. Section 5.5) illustrates the conceptual difficulty concerning fixed capital goods, which has to be overcome in the construction of an intertemporal production function.[22] The period t should be regarded as a weighted average of as many production periods as there are capital goods in use. Whereas the ideal production period for each capital good is a purely technical characteristic, it is difficult in the determination of the average length of the production period to avoid a weighting based on an economic valuation, where, for example, the rate of interest will play a role.[23] The average production period then represents a tangled skein of technical possibilities and economic calculations and is no longer suitable for characterizing the technical aspects of production.[24] Nor is it then possible to attach an unambiguous meaning to changes in the length of the production period, since changes in the technology are not separated from substitution dictated by economic considerations. 'The Austrian theory thus ignores any process of change'.[25] We must therefore conclude that any attempt to describe an average production period as a price-independent measure of capital and as a characteristic of the technique of production along the lines of the Austrian theory of capital fails, because different lengths of the production periods of the factors of production cannot be aggregated without expressing the economic significance of the various periods in terms of prices.[26]

This conclusion does not mean that the concept of an intertemporal production function should be abandoned, but simply that the construction of an average production period as a technical concept must be rejected.[27] If we want to show that the amount x of a good produced in period T depends on the factors of production used in earlier periods, the obvious thing to do is to introduce separately the dated inputs of the various factors of production. If there are n factors of production and if their quantities in periods t are denoted by $v_{1t}, \ldots, v_{nt}$, then the intertemporal production function can be written in the form:

$$x_T = f(v_{1t}, \ldots, v_{nt}, v_{1,t+1}, \ldots, v_{n,t+1}, \ldots, v_{1T}, \ldots, v_{nT})$$

This form of production function, which gives an unambiguous description of the state of technology, shows that it is impossible to operate with a mean production period without either lumping together dissimilar magnitudes or introducing prices. An important point is that such an aggregating operation is in fact not necessary for determining the optimum structure of production in a firm whose aim is to maximize profits.

The intertemporal production function expressed as a relationship between output at a given time and the capital goods used in the production process in various earlier periods, is clearly distinguished from formulations such as Wicksell's, in which a direct connection is postulated between output and the average production period. In Hicks' words, 'A production process must now be defined as a stream of inputs, giving rise to a stream of outputs'.[28] It is also important to distinguish the intertemporal production function from dynamic production functions, which describe the development of production as a function of the factors of production used in each period, and in which the production in each period is treated as an instantantaneous process.[29] Nor should Salter's production function, in which the lifespan of capital goods is taken as an independent variable, be confused with an intertemporal production function.[30]

Describing the state of technology with the aid of an intertemporal production function (of which, as in the case of the non-intertemporal production functions, further variants can also be formulated in the form, for example, of the Cobb-Douglas and the CES functions), the main manifestation of technical development is its influence on the intertemporal nature of production. An increase in technical knowledge may mean that other connections are discovered between the factors of production in current use and the expected level of production at a later date. A new intertemporal production function then appears, which—like the non-intertemporal ones—can be modified by the creation of new products and a change in the nature of the factors of production.

8.5 TECHNICAL POSSIBILITIES

The description of technical possibilities by the production function shows a certain diversity. Different assumptions are made about the extent of substitution between the factors of production, and diminishing, constant and increasing returns to scale can in principle be taken into consideration. In some cases the production functions reflect increasing, diminishing and constant returns in different intervals, while in other cases they do not. The fact that production is a process in time can be expressed.

It is tempting to examine the significance of these characteristics of the production process, but such an attempt is hampered by the conceptual difficulty that the production function, regarded as an *ex ante* description of the technical possibilities of the enterprise, cannot be observed directly. The actual choice of certain combinations of means of production depends for example on the objectives of the firm and the prices of the means of production. The changes observed in the use of the means of production reflect a shift in the equilibrium position, and only in very exceptional cases do they give an idea of some points along the production function.[31] It is admittedly possible to discover directly what pattern the relationship between production and the factors of production follows in the course of time, but this is not an observation of the production function in the sense that is intended here. In this situation discussions of the empirical characteristics of the production function are somewhat speculative. Economic theory copes with this situation by formulating production functions that are as general as possible, i.e. production functions that can accommodate several variants. The *a priori* exclusion of certain possibilities often condemns empirical investigations to failure, which emphasizes the value of generalized production functions.

Apart from the construction of general production functions, in which no comment is made *a priori* about a number of aspects, there is another approach in economic theory. Production functions indicate the maximum possible output for each combination of the means of production, and therefore rely on a technical maximization process. For this reason, they are not derived from the technical possibilities, but are instead postulated. The characteristics of the production function are however not independent of the character of the set of all the technical possibilities.[32] This character can be expressed by means of a set of postulates, and leads to the development of a production theory based on the theory of sets.

Such a production theory, which gives a complete description of all the technical possibilities, has the added advantage of providing an elegant way of accounting for some other aspects of the production technology, such as the joint production of several different goods. Further, as Frank has demonstrated, this theory is very suitable for dealing with the problem of the indivisibility of the means of production.[33] Besides—and this is particularly important in the present context—it gives a description of the technical possibilities that is not based on aggregation.

Economists tend to use the aggregation method in the construction of production functions either by including capital and labour in their entirety or by introducing average production periods. The result is that the production function can no longer be considered a pure description of efficient technical possibilities, since prices must be introduced to establish a common denominator for capital goods and the various types of labour. Admittedly, this difficulty can also be overcome without the set-theoretical production theory, by expressing the different means of production in an explicit form, as has already been illustrated in the case of the intertemporal production function, but this function must then itself be derived from a series of hypotheses concerning the technique of the production process. The introduction of these postulates therefore changes not only the form of the theory, but also its content. However, it is still possible to examine the relationship between the production function as the starting point of the traditional production theory and the set of production possibilities, which is the basis of the modern production theory. Jorgenson and Lau[34] have reversed the conventional line of reasoning and derived some properties of the set of production possibilities from those of the production functions.

8.6 SET-THEORETICAL PRODUCTION THEORY

The technical possibilities open to an enterprise form a set of possibilities. Each technical possibility is called a process or activity, and consists of the creation of goods by the consumption of other goods—the means of production. Only special goods, which can be both means of production and end-products, are included here, so that strictly speaking the only important point is whether a good is introduced into a process or results from it. For the sake of simplicity, however, we shall continue using the terms 'means of production' and 'end-products'.

The possible production processes that can be used are the elements of the set, which will be called the 'production set'. If a certain quantity of a good is used up in an activity, it is given a negative sign, while a quantity of a good made in the production process is given a positive sign. Each activity is then described by a row of numbers, which can include zero when the good in question does not take part in the process.[35] The length of the row of numbers depends on the number of goods used for characterizing the production process. Each number in the row is regarded as the component of a vector. In other words, the production set is composed of a number of vectors, each of which gives a full quantitative description of a process, and the total number of which corresponds to the total number of processes or technical possibilities. If n goods are included, the production set is a sub-set of n-dimensional Euclidean space, in which each coordinate corresponds to a certain good.

A number of postulates will now be formulated about the production set, which collectively express the assumptions concerning the technical possibilities. These postulates are not binding and can of course be replaced by others; however, only the most common hypotheses will be discussed here.[36]

It is assumed first of all that the boundary of the production set is included in the set. This assumption is essential for the economic considerations that follow from the construction of the set of technical possibilities. The choice of a production process that is optimal for example from the point of view of profits always involves a boundary point of the production set, since this is formed by the efficient processes. Secondly, it is assumed that there are some vectors in which positive numbers occur and some in which they do not. This means that there are some technical possibilities in which goods are only inputs and not outputs, and there are some processes in which goods *are* produced.

Another important hypothesis is that production is irreversible, i.e. if the technical possibility $(-3, -5, +6)$ occurs, then the technical possibility $(+3, +5, -6)$ does not exist. One process cannot therefore cancel out the other. Together with the hypothesis that production processes in which there is no output do exist, this postulate excludes the possibility of processes in which an end-product is produced, without any good being used up.

Having this characterized each technical possibility by a certain combination of input goods and output goods, we can now introduce the scale of production. The multiplication of all the components of the vector v by a non-negative number λ gives a new vector λv that corresponds to a scale of production which is λ times as large as the scale of the original activity. The interesting question now is whether this new process belongs to the technical possibilities of the firm. We must look at the three situations of constant, decreasing and increasing returns to scale.

Constant returns to scale obtain when, for each process

included in the production set, the multiplication by λ (≥ 0) gives a technical possibility for the firm. The resulting possibilities therefore all belong to the production set. If we want to speak of constant returns with all the technical possibilities, then this characteristic should apply to all the activities. Decreasing returns to scale obtain when, for each activity included in the production set, the scale of production can be arbitrarily reduced, while an increase in the scale of efficient activities is impossible because this would be outside the set of technical possibilities. Increasing returns to scale obtain when it is possible to increase arbitrarily the scale of each activity belonging to the production set, whereas a decrease in the scale of an efficient activity takes us outside the production set.[37]

The case of constant returns to scale, which has figured largely in economic debate, is mostly equated with what is called 'linear technology' in the set-theoretical production theory. Such a linear technology can be postulated, but it can also be derived from some other assumptions. We shall consider in this connection two technical possibilities belonging to the production set. The production set is said to be additive if the addition of any two technical possibilities gives a new technical possibility that belongs to the production set. The repetition of an activity is a special case. The additivity postulate may however be at loggerheads with the limited quantity of one or more means of production.[38] It may be that this postulate holds for a sector of industry but not for an individual firm.

The additivity of the activities does not necessarily mean that the goods are divisible. To postulate the divisibility of goods, each pair of activities must fulfil the requirement that the vectors lying in between should also represent technical possibilities that belong to the production set. If this requirement is fulfilled, the production set is said to be convex. These two postulates, combined with the hypothesis that an activity with zero production is also possible, imply the case of constant returns to scale.

We have so far outlined technical possibilities and represented them in a truly microeconomic manner. Aggregation problems have not occurred, since we have always assumed specific goods. The microeconomic production function can now be derived by examining what quantities of end-products are involved in the efficient activities constituting the boundary of the production set. The characteristics of the production function are very closely connected with the hypotheses introduced in connection with the production set.[39] Thus, if the production set is such that constant returns to scale arise, this characteristic will also be exhibited by the production function. Whether the production function indicates substitution between the means of production depends on whether efficient activities have been included in the set of technical possibilities, where the same quantity of product is obtained with various quantities of the means of production. Otherwise we are confronted with limitational relationships, and this circumstance leads to fixed technical coefficients in the case of linear technology.

The concept of substitution used here is more general than that used in current theories, since the possibility of substitution need not cover activities lying close to each other. The typical production theory usually regards substitution in such a way that the criterion mentioned above applies to a continuous series of efficient technical possibilities. Axiomatic production theory affords a very elegant solution of the problems of whether the technology involved in the production process is better described by Leontief's production function or by the production functions which acknowledge the possibility of mutual substitution between the means of production, as exemplified by the Cobb-Douglas and the CES function. It advocates, by way of synthesis, that, for each activity, the ratios between the means of production and the end-products are fixed, but the substitutability of the means of production in the production function varies with the number of activities that have the same quantity of end-product and different quantities of means of production in the set of efficient activities. Against this background, Leontief's production functions and the CES production functions can be regarded as extreme representations of the technical possibilities. However, they are generally more amenable to a mathematical treatment than the production functions based on production sets that correspond to reality.

8.7 STATE OF TECHNOLOGY AND THE PRODUCTION FUNCTION

This brief description of set-theoretical production theory has shown that all the technical possibilities open to the firm in a given period of time can be described in a systematic and detailed manner. The production function that can be derived from the production set gives the efficient combinations of the means of production which imply a technically optimal organization of production.

The next question is whether the production function reflects the state of technology available to the firm in a given period of time.[40] The previous considerations indicate that this cannot be taken for granted. The set of all the technical possibilities from which the production function is derived depends not only on the state of technology in the sense of technical knowledge available to the enterprise; the possible combinations of the means of production and the nature and volume of the corresponding output depend also for example on the quality of the available factors of production, i.e. labour and capital, and the extent to which division of labour is possible within the firm. Education, the diffusion of technical knowledge, the employees' living conditions, environmental hygiene, laws of all kinds, and the size of the firm are thus involved in the production set. In short, therefore, the technical possibilities portrayed by the production function depend not only on the available technical knowledge, but also on all the circumstances that affect the potential combinations of the means of production in a given period. Conversely, it can be said that the state of technology in the strict sense is depicted by the production function.

If the state of technology is construed in a broad sense, so that all the technical possibilities are encompassed by it, then it is fully expressed by the production function; if, on the other hand, it is construed in the narrow sense, i.e. as meaning the available technical knowledge, then it can no longer be said that the production function reflects the state of technology exclusively, for the function then reflects all the technical possibilities determined not only by technical knowledge but also by the other conditions of production.

One might conclude that the production function is a purely

technical relationship, since all the conditions mentioned converge in the set of today's efficient and technically feasible combinations. In the framework of a static model, therefore, the production function can be regarded as an exogenously given relationship from which other economic characteristics can be explained. However, a production function also has a past: today's production function is partly the result of a series of decisions made in the past, which almost always relate to the use of scarce resources that can be used as alternatives, and to this extent it is the product of past economic decisions. In the framework of a dynamic analysis, one must explain today's production function partly in terms of previous economic decisions; the significance of this for economic theory is that the production function is at least partly an endogenous relation. The extent to which this is the case depends on the importance attributed to the autonomous component in the expansion of technical knowledge. Even though nowadays the progressive trend of technical knowledge and technical possibilities are to a great extent the result of deliberate economic decisions, inventions are still being made that do not arise from a conscious and specific use of scarce resources.

To conclude, economics should abandon the view that the production function is a relation given in an exclusively exogeneous manner. As the expansion of technical knowledge and technical possibilities becomes more dependent on the use of scarce resources, the endogenous character of the production function increases, so that economists should consider it one of their tasks to explain the production function on the basis of economic parameters. A corollary of this is that, in the formulation of economic policies, technical development should no longer be treated as a process that is in principle not amenable to control.

8.8 TECHNICAL DEVELOPMENT AND THE PRODUCTION FUNCTION

Production functions have not only a past, but also a future. For the firm the production function of tomorrow generally has a greater scope than the present one. Can the appearance of a new production function be identified with technical development? If the latter is taken to mean nothing more than the increase of technical knowledge (technical development in the narrow sense), then the answer must be simply 'no', for a new production function can also be due to an improvement in the quality of labour as a result of education and training. If, however, technical development is taken to mean increase and change in the technical possibilities, then by definition a new production function must be identified with technical development.

Various reasons have encouraged us to distinguish between a broad and a narrow meaning of the concept of technical development. Technical development in the broad and narrow sense behave to some extent as effect and cause, but it should be borne in mind that there are also other causes of technical development in the broad sense, such as changes in the living conditions of the labour force, improvement of the environment and changes in the size of the firm. Furthermore, the technical possibilities affect the production of new technical knowledge.

On the other hand, does technical development lead to a new production function? The answer must be 'yes', both for the broad and narrow interpretation of technical development. It is widely held that technical change can be regarded as a shift in the production function. Here the starting point is that the new production function arises from the old as a result of one or more of the coefficients being changed. It is not impossible according to either interpretation of technical change that a new production function can be derived from the old one by altering one or more of its parameters while keeping its form intact. As regards technical development in the narrow sense, we can think here of an increase in technical knowledge that causes not a qualitative but a quantitative change in the relationship between production and its factors. Technical development in the broad sense can also be such that the form of the production function does not change with time, only its coefficients do.

However, we cannot stop here. Our account of technical development in the twentieth century has shown that technical knowledge often undergoes fundamental changes at the level of the firm. Essential changes come about in the production set through the creation of new means of production, new products and new relationships between the goods. These changes have an effect on the production function derived from the production set. New production functions arise which differ from the previous ones in a discontinuous manner. The custom of aggregating capital goods and various types of labour in the construction of the microeconomic production function may have led to the misunderstanding that technical development can be generally characterized as a smooth continuous process. Even when the broad interpretation of technical change is used, and thus attention is focussed on new technical possibilities, it must be concluded that new production functions arise on account of radical changes in the social structure, as a result of which the nature and number of technical possibilities for the firm undergo fundamental changes.

Not only is technical change interpreted in economic literature as a shift in the production function, but, conversely, a shift in the production function is often taken as a sign of technical change. What economists quite often have in mind is a simple shift in the production function, with a larger output from the same combination of means of production in a subsequent period of time. A causal connection is then assured between the production functions for different periods. These economists generally assume technical change in the narrow sense. It can now be seen that a shift in the production function can indeed be due to technical change in the narrow sense, but we must reject a simple connection between such a shift and technical change. This is partly because the production function may be shifted in the same way by factors that lie outside technical development in the narrow sense (so that technical development in the broad sense is then the relevant concept), and partly because this attitude does not take into account the diversity of the causes and effects of technical development. Even when the shift of the production function is taken in such a wide sense that it encompasses any change in one or more of the parameters of the function, the view of technical development still remains restricted by the formulation of the production function, which is the starting point of the argument under review. The production of entirely new goods and the expansion of technical possibilities by the construction of entirely new means of production lie quite outside the scope of this argument.

We can see so far that an incomplete picture is given of technical development by expressing it as a shift in the production function, for an increase in technical knowledge and the creation of new technical possibilities cannot generally be squeezed into a shifting production function, and the diversity of technical development is better expressed by identifying the latter with the appearance of new production functions.[41]

It is interesting to compare this conclusion with the frequently voiced, and apparently contradictory, view that technical development can be thought of as a gradual and continuous process in time. This view is not contradictory because in the first place the actual course of technical development depends not only on the expansion of technical possibilities that are described by the production function, but also on the choice the entrepreneurs make from the wealth of potentially useful techniques, and the gradual application of new technical possibilities should not be confused with a gradually shifting production function. But even if the actual course of technical development is considered to be gradual and continuous, we must still reject the notion of an invariant causal connection between the production functions for different periods. The gradual creation of new products and new means of production does not detract from the qualitative character of the change, which is not described adequately by a shifting production function, since this only accommodates quantitative changes.

8.9 TECHNICAL DEVELOPMENT AND SET-THEORETICAL PRODUCTION THEORY

The above argument can be further elucidated by examining the concept of the new production function. An objection could be raised to this argument that the difference between the old and the new production function is not sufficiently sharp, especially when the shift in the production function is considered in such a broad sense that it encompasses an arbitrary change in one or more of the coefficients. If the production function is formulated in a sufficiently general form, then the new set of coefficients is always in a certain relationship to the old set, while the form of the function remains the same.

In this light, the transition from a Cobb-Douglas production function to a CES function must be regarded as the appearance of a new production function, unless for example a sufficiently general VES function is postulated from the outset. A change in the elasticity of substitution is then a quantitative change in a parameter, as a result of which we are faced with a shift in the general production function. Thus each new manifestation is as it were imminent in the previous one, provided that this has been described in sufficiently general terms.

Having arrived at this point, we can now profitably return to the set-theoretical production theory from which the production function has been derived. The set of all the technical possibilities in each period is accounted for in a detailed manner in this theory, and—what is particularly important—all the goods are mentioned separately. Each activity is characterized by a quantitative relation between production and the means of production. The characteristics of the production function that can be derived from the production set are partly determined by the postulates introduced. The meaning of a shift in the production function based on the production set is invariably that the efficient activities undergo a quantitative change. We are then concerned with changes in the parameters which are either accommodated by the design of the old production function, or can be derived indirectly from it by reformulating it in the light of the new production function. In this sense, the new production can always be derived from the old one, and one can speak of a shift in the production function. This quantitative change in the efficient activities expresses a change in technical knowledge, which can be regarded as a shift in the production function.[42]

We have seen before, however, that technical development comprises in particular the creation of new products and new means of production. The production set is then considerably enlarged by the introduction of entirely new processes, and so an entirely new production set is created. The production function derived from the new production set is essentially different from the old one, since it comprises new products and new means of production. Thus technical development manifests itself in the appearance of new production functions that cannot be derived from the previous one, because the underlying production sets differ in the nature and number of goods. This shows the great value of the set-theoretical production theory for the analysis of technical development.[43]

The investigation can now turn to the new production sets and their properties. It is not impossible that this will reveal forms of technical development characterized by a temporal consecutiveness that is more complicated than has so far been imagined by economists.[44] The difficulty is that in so far as the new production sets are due to scarce resources becoming available, it still cannot be predicted what new products will be created and what new capital goods will be necessary for production, apart from the fact that inventions that renew the production sets will also be made independently of the economic process. Given all this, the production set and its construction should be the object, rather than the starting point, of the analysis in economic theory. Although, in the framework of a static analysis, production sets can be regarded as technical data, the change of one set into the other in the course of time is so strongly influenced by economic factors that the dynamic analysis of technical development cannot ignore the economic effects on the formation of new production sets.

8.10 OTHER VIEWS

To put these ideas in a wider context, we will now examine some other theories. Brown,[45] for example, mentions four characteristics of the production function: the efficiency of the technology, the existence of returns to scale, the capital intensity, and the extent to which labour and capital can be substituted for each other. He calls these four characteristics an 'abstract technology', and expresses technical development as changes in this abstract technology. The theoretical framework of the production function (the formulation of which can be more or less refined) is retained once it has been postulated, so that the analysis is limited to the aspects of technical development that can be reduced to a shift in the production function.

Objections to this excessively narrow view from the standpoint of business economics have been discussed by Schätzle,[46] who believes that the 'technical horizon' of the firm cannot be expressed by a production function since only qualitatively equivalent goods can be aggregated into a factor of production. Besides, there is a different production function for each additional type of good. Schätzle therefore rejects the assumption that technical development can be regarded as a shift in the production function. Even though we endorse his conclusion, his reasoning is unconvincing. It is difficult to see why one should distinguish as many production functions as there are production methods. With the aid of the set-theoretical production theory, in which no aggregation is in fact carried out, we can derive from the complete description of all the technical possibilities a production function in which the relationships between all the goods—end-products and means of production—are expressed simultaneously. Schätzle's conclusion that technical development at the level of the firm produces the number of production functions is therefore unsatisfactory.

In any case, the analysis of the creation of new technical possibilities should be separated from the analysis of the entrepreneur's choice of optimal production technique, in which profits and cost calculations definitely play a part. Even though it has been assumed that the increase in technical knowledge is to a considerable extent determined endogenously, and thus calls also for an economic explanation, the latter is so different from the economic analysis of the choice of optimal production technique that it is desirable to keep the two problems separate.

8.11 CONCLUSIONS

Microeconomic production theory is a good illustration of the custom of choosing the available technical possibilities as a starting point for economic analysis. This is often expressed by saying that the state of technology can be taken as a non-economic factor, i.e. as one of the given data of economic science. In the light of the preceding considerations, this definition is restrictive, since technical knowledge has been included amongst the technical possibilities. Thus, it has been seen that education for example affects the technical possibilities because it influences labour. It has an effect both on the level and on the nature of technical knowledge, and, conversely, the latter affects the method and content of education. There are many more connections between the various components that determine the technical possibilities, so that technical development in the narrow and the broad sense are not independent of each other either. In this sense, we can endorse Schumpeter's view that the opening-up of a new market gives a new production function, provided that it results in the internal growth of the firm, which is accompanied by new technical possibilities.

Both technical knowledge and technical possibilities are to some extent dependent on the use of scarce resources. It is therefore becoming increasingly difficult to treat technical knowledge and the range of technical possibilities as non-economic factors. It need not concern us here that the technical possibilities depend on other things—such as the quality of the factors of production—which are often included in the other non-economic factors.

In the most common version of microeconomic production theory, the technical possibilities are portrayed by the production function. This production function is taken as a starting point for the economic analysis of the behaviour of the firm, and is not explained any further by analysis of economic considerations. The production function is considered to be a technical relationship that describes the efficient combinations of the means of production. The technical nature of the production function has been questioned on the grounds that calling certain combinations efficient implies that a choice has already been made between scarce inputs, so that the choice of an efficient rather than an inefficient combination relies on an economic judgment.

Since this does not permit any distinction between the technical and the economic approach, this way of looking at things is unhelpful. The selection of the most efficient combinations takes place without any reference to the prices of the factors of production and is, in this sense, a technical choice. This does not mean there is no wastage due to the use of a larger amount of a factor than is technically necessary for a given output. While this implies an economic wastage as well, it does not deprive the production function of its technical character.

A similar conclusion results where a government restricts the set of technical possibilities. It is possible that the government might for some reason require the use either of more labour or of another means of production than is technically necessary. The combinations that would be efficient without government intervention are therefore unattainable for the firm, and the production function should then be based on the efficient combinations that are attainable. The technical maximizing process in itself does not brand the production function as an endogenous relation that has to be explained by economics.

Examination of technical development in the narrow sense and in the broad one has changed our perspective in several ways. The increase in the technical possibilities in a dynamic framework depends also on economic decisions about the nature and trend of investment in labour and capital goods. The state of technology can no longer be treated as a non-economic factor, because it is too dependent on economic processes. Economic science can no longer avoid bringing technology and technical development within its boundaries. The endogenous character of the production function is a feature of the general belief that the whole field of technical possibilities is to be treated as something that needs, rather than offers, an explanation.

The belief that it follows from this that the production function can now no longer be considered a technical relation results from a misunderstanding. The dependence of the production function on previous economic decisions does not mean that the relationships between the means of production and the end products are no longer of a purely technical nature. This confusion, sometimes encountered in economic thought, can be ascribed to the aggregation problem, which affects both the microeconomic and the macroeconomic production function. This is because the production function cannot be regarded as a purely technical relation if the aggregation of the capital goods, the various types of labour and the end products is expressed by introducing the prices of these different goods.[47] The idea that the boundary between economic choice and technical conditions is extremely blurred here may also arise from the application of a certain technical method, which is partly dependent on relative prices and through which new technical possibilities can arise.[48]

This interdependence between inventions and innovations, as a result of which a steady stream of technical changes can actually be observed in the course of time, has prompted some economists who have rejected the view that a shift in the microeconomic production function is a criterion of technical development to accept that 'aggregate technological change can be represented by a smooth time trend'.[49] This hypothesis is the basis of the frequent practice of characterizing technical development in the macroeconomic production function by the introduction of time as an independent variable.

9 Macroeconomic production theory

Ever since the beginning of economics, macro and micro theory have existed side by side; they will continue to do so in the future. Each is needed, neither is expendable.

F. Machlup[1]

9.1 INTRODUCTION

Although the explanation of economic phenomena should ultimately be based on microeconomic behaviour, macroeconomic analysis cannot be ignored[2]; this is partly because some macroeconomic background is needed before we can assess the relationship between technical development and economic growth in the next chapter, and partly because the arguments about the existence of a macroeconomic production function are relevant to the interpretation of the significance of macroeconomic growth models.

Highlighting the derivation of macroeconomic equilibria can distract one's attention from the fact that by doing this one ignores the complex microeconomic relationships. This is particularly true when the macroeconomic connections are not derived from hypotheses about microeconomic behaviour but are simply postulated. A macroeconomic production function which is introduced by analogy with a microeconomic one without discussing the connection between them produces only a theoretical analysis. In microeconomics the hypothesis of constant returns to scale is connected with a change in the size of the firm, but in macroeconomics it only means a certain increase in the factors of production labour and capital.

Nevertheless, a macroeconomic analysis—even without a solid microeconomic basis—may be considered a provisional approximation. In Chapter 10 we re-encounter the compensation theory (discussed in Chapter 2) when analysing the effect of technical development on the labour market. Its relevance to macroeconomics turns on whether sufficient compensating employment is created in the economy at large to re-employ those who, through technical development, have lost their jobs.

Considerable empirical research which is relevant to the macroeconomic analysis of growth has been carried out, and at a later stage we shall examine the significance of this research in detail.

Finally, macroeconomic analysis has been instrumental in the formulation of overall economic policy. Microeconomic and macroeconomic approaches complement each other in this field, but first we need a rough sketch of the effect of technical development on society in general. Central governments not infrequently trace out a policy on this basis alone.[3]

When the scales in economic theory, empirical research or economic policy tip too much in favour of the macroeconomic approach, there is usually a microeconomic reaction. Mention should here be made here of Salter's important theoretical and empirical work on technical change, which is significantly different from most treatments of this subject based on a very high degree of aggregation. In the following chapters we will try to keep a balance between these two complementary approaches.

9.2 THE EXISTENCE OF A MACROECONOMIC PRODUCTION FUNCTION

The aggregation problem concerning the stock of capital goods appears on both the microeconomic and the macroeconomic level: in the former, the problem is mainly the aggregation of heterogeneous capital goods, used by the firm, into its capital stock,[4] while on the macroeconomic level there is the added problem of aggregation over all firms. Aggregations of these and other types have been dealt with by K. Sato in an important book on production functions and aggregation.[5]

The possibility of aggregating heterogeneous capital goods without the use of relative prices has been seriously questioned by Joan Robinson[6] in particular. Closely connected with this is her criticism that in neo-classical economics the distribution is determined by the marginal products derived from a production function which presupposes a knowledge of the factor prices. It is surprising that the discussion is mostly restricted to the difficulties of aggregating the capital goods and does not deal with the analogous problem of aggregating different types of labour and heterogeneous end-products.

Even in a greatly simplified example, where only one capital good is used by each enterprise, all the enterprises use the same type of labour and a homogeneous end-product is obtained, the aggregation of the microeconomic production functions into a corresponding macroeconomic function is possible only if the individual production functions can be resolved into two parts, one for the use of labour and the other for the use of capital. The identical parts can then be lumped together.[7]

Fisher[8] has pointed out that the aggregation problem has not been presented in an entirely correct manner here, since no use has been made of the fact that a production function reflects the efficient combinations. The existence of a macroeconomic stock of capital goods can be demonstrated[9] if we assume (a) an optimal allocation of the homogeneous labour, (b) constant returns to scale, and (c) a neutral technical change in Solow's sense.[10] Whitaker has examined the question whether the capital goods can generally be aggregated more successfully if one starts with an optimal allocation of labour[11], and his answer is predominantly in the negative.

Although Solow fully realizes that a macroeconomic production function can be extracted from the wide variety of microeconomic situations only if greatly limiting constraints are imposed, he holds that such a macroeconomic production function must be postulated by way of an all-embracing description, in order to make it possible to carry out a series of macroeconomic investigations of economic growth.[12] Samuelson, while recognizing the great importance of Solow's empirical work, does not find this entirely satisfactory and has tried to give Solow's method a sounder theoretical basis by constructing a so-called 'surrogate' production function. He shows, in particular, that the conclusions that can be drawn from heterogeneous capital models of the general equilibrium type agree with the conclusions that can be drawn from a surrogate production function which includes labour and the surrogate capital. This surrogate function is thus a substitute for the complicated and extensive description of all the technical possibilities.

Samuelson[13] uses a two-sector model in which a finite number of heterogeneous capital goods are distinguished in such a way that only one capital good is included for each technique, and in which Leontief's production functions are utilized. With the aid of this model, he demonstrates that, in the case of identical factor ratios (i.e. identical ratios between the factors of production) in the consumer goods and capital goods sectors, the same conclusions can be drawn as from the simple neo-classical approach, in which capital is regarded as a homogeneous entity, the returns to scale are constant and the equilibrium values are ensured by perfect competition. For these values, we have (a) a positive correlation between the real wage rate and the ratio between labour and capital, (b) a negative correlation between the rate of interest and this ratio and (c) a lower rate of interest corresponding to a higher real wage rate.

The case involving two techniques is illustrated graphically in Figure 5. In his model of Marx's theory, published in 1957, Samuelson discussed the characteristics of the wage-interest frontier, which—both in that model and in the one published in 1962—is the envelope of the linear wage-interest curves that can be constructed for each technique. The wage-interest frontier is obtained by examining which of the technical possibilities produces the highest wage rate at any given rate of interest. The characteristics of the envelope are consistent with the results obtained with simple neo-classical assumptions. The wage-interest frontier has a negative slope, so that, the technical possibilities being given, a lower rate of interest corresponds to a higher real wage rate. Furthermore, in the case of straight wage-interest curves, considered here, the frontier is convex towards the origin of the coordinate system, so that a technique that is efficient in a certain range of wage rates does not become efficient again after another technique has become the most efficient one. In the case of two techniques α and β, shown in Figure 5, technique α is efficient only for values of r greater than OB, while technique β is efficient only for values of r smaller than OB. An essential point is the linearity of the wage-interest curves, which is due to the identity of the factor ratios and to the

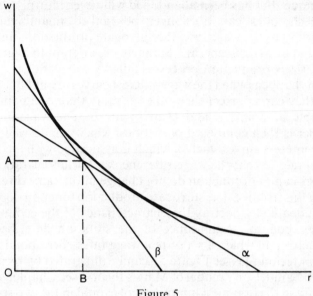

Figure 5

fact that no wage discounting is carried out. Thus a certain technique belongs to each set of efficient wage rates, and the techniques can be ranked unambiguously in accordance with these wage rates. The surrogate production function of the neo-classical type is constructed precisely on the basis that a single hypothetical technique that is characterized by a certain ratio between labour and surrogate capital is taken to represent every real technique that is characterized by a certain ratio between labour and capital. The wage-interest frontier of the substitute is then exactly the same as that in the extended model constructed by Samuelson, in which each technique is characterized by just one capital good. A movement along the wage-interest frontier then implies that the wage rate and the output per head change in the same direction.

If the set of technical possibilities is so constructed in reality that the same technique is efficient at different wage levels, separated from one another by wage values at which another technique is efficient, the surrogate production function can no longer serve as an acceptable substitute model for this economy, since according to this surrogate function a technique that has been abandoned will never return.[14] When the techniques can no longer be ranked unambiguously according to the wage rate, the surrogate production function vanishes from the scene, and Samuelson's attempt to aggregate the various production processes fails.[15]

This happens when the wage-interest curves are not linear, so that they can intersect each other at more than one point, for example at points P and Q in Figure 6. The wage-interest frontier is then composed of sections, which are derived from wage-interest curves, each of which is associated with a certain technique, so that the wage rate, the capital/labour ratio and the per-capita production do not change in the same direction along this frontier. The surrogate production function does not then adequately portray the characteristics of the economy.

The economic significance of the situation described by Samuelson, in which a set of real wage rates corresponds to a certain technique (see Figure 5), can be illustrated by recalling that in Samuelson's model of Marx's theory (see Chapter 3) the identity of the organic composition of capital in the two sectors meant that $a_1 b_2 = a_2 b_1$, which can also be written as $a_1/b_1 = a_2/b_2$; in other words the ratio between the factors of

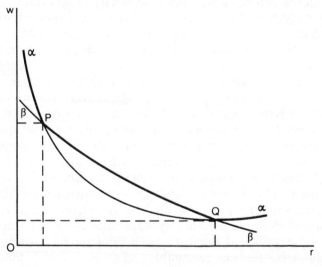

Figure 6

production is the same as in the two sectors. As we have seen, this is the essential condition for the linearity of the wage-interest curves, and thus for the construction of the surrogate production function. If the organic composition of capital is different in the two sectors, then the wage-interest curves are not linear, and it is not possible to rank the technical possibilities unambiguously according to the level of the real wage,[16] because previous techniques return.

It can be shown that in the case of linear wage-interest curves the price of capital goods p is independent of the rate of interest r and depends only on the set of technical coefficients describing the technical features of the economy under consideration. The price p can then be used as a yardstick in the aggregation process. If, however, the condition for linear wage-interest curves is not fulfilled, the price of capital goods is generally a function of the rate of interest, so that the stock of capital is no longer an independent and purely technical quantity.

The possibility of a 'perverse relationship' between the real wage rate and the capital/labour ratio, known in the literature as 'Ruth Cohen's curiosum', was recognized by Joan Robinson, who came to the conclusion that it could be ignored because it was an exception. Robinson generally assumed that a higher degree of mechanization corresponds to a higher real wage.[17] However, Sraffa demonstrated in his *Production of Commodities* that the impossibility of ranking various techniques unambiguously according to the level of real wages is the rule rather than the exception.[18] Joan Robinson subsequently used Sraffa's work in order to continue her attack on the macroeconomic production function and on the aggregation underlying it.[19]

Sraffa's general conclusion was attacked in 1965 by Levhari,[20] who believed he was able to demonstrate that it applied only to one branch of industry and not to the whole economy. As Pasinetti[21] has shown, however, Levhari's proof is unsound, and Samuelson and Levhari subsequently admitted in a joint publication[22] that Sraffa's conclusion does in fact have general validity.

As Samuelson himself shows, the significance of the return of techniques with an increase in the real wage or a decrease in the rate of interest is that the contention of Jevons, Böhm-Bawerk and Wicksell, in which a lower interest rate is associated with the application of more highly mechanized techniques (i.e. a longer roundabout route of production), is no longer tenable.[23] As regards Wicksell, the cautious view has so far been proposed that the return of techniques can be traced back to the Wicksell effect.[24] In this connection, we can distinguish between 'real' and 'price' Wicksell effects: the former is exerted on the value of capital by a change in the wage rate w or in the rate of interest r along a wage-interest curve that corresponds to one and the same technique, while the second is exerted on the value of capital by a change in the wage rate w or in the rate of interest r when this change leads to a change in technique, so that we move along the wage-interest frontier. The real Wicksell effect generally embraces the price Wicksell effect and a change in production techniques.

If the wage-interest curves are linear, there is no price Wicksell effect, for the price of capital goods is independent of the rate of interest r, and the real Wicksell effect then reflects only a change in techniques. In other, more complicated cases, it depends on the sign of the slope of the wage-interest curves whether the Wicksell effects are positive or negative. In general, therefore, Wicksell did not think in terms of a simple time pattern of the inputs of production.

If the quantities of the factors of production applied do not decrease uniformly over the course of time as the production period is extended, then the average production time is no longer determined uniquely, and a previously abandoned technique can again become efficient as a result of changes in the rate of interest in the same direction. In the non-temporal production model, techniques can return if more capital goods are assumed per technique,[25] or if the ratio between the factors of production is not the same in the capital goods sector as in the consumer goods sector. After giving a penetrating summary of the 're-switching debate', Wan therefore concludes that it does not make any difference if the capital goods are homogeneous in the physical sense, for the question is how heterogeneous they are in the economic sense.[26] This heterogeneity refers to the various ways in which the same capital goods can be produced, and to the various services which the same capital goods can render in the production process.

It also appears from Spaventa's study[27] of a simple production model that the question is not so much the heterogeneity of capital in the case of various techniques as the heterogeneity of the processes, which is reflected in differences in capital intensity. Ayküz has argued that the condition for the reswitching of techniques also depends on the measure in terms of which the trade-off between the real wage rate and the rate of profit is expressed,[28] but Jaeger has proved that this view is incorrect.[29] However, the debate about the conditions required for reswitching continues.[30]

If, irrespective of the heterogeneity of the capital goods, we assume that the ratios in which the factors of production are used are the same in all the sectors for each technique, then all the essential variables behave as if there were only one good involved. If this hypothesis is abandoned—as is unavoidable when technical development is being examined—then the unlike scarcity ratios exert a decisive influence on the results of the analysis, particularly as regards the construction of a macroeconomic production function.[31] To repeat, the important point is therefore not heterogeneity in the technical sense but in the economic sense.[32] The difficulty involved here is really two-fold: (1) the possibility of reswitching hampers the construction of macroeconomic production function, and (2) the price of capital goods is a function of the rate of interest (or profit) r in the general case. If we look back in time, we see that a capital good is a produced good and its value also depends on the time structure of its production, so that r enters into its price. If we look forward in time, we see that the price of the capital good depends on the service the good may render, and is therefore also dependent on the rate of interest. Blaug[33] seems to overlook the crucial role of the rate of profit in the valuation of capital in his defence of the neo-classical position. The rate of profit is not a consequence of the production side of the economy, but instead the production structure depends on the exogenously determined rate of profits.[34] One could even say that the reswitching phenomenon is unimportant for the outcome of the Cambridge controversy.[35] As for the time structure of the production of capital goods, there is a difference between capital and labour in so far as labour is treated as an original factor of production. If labour is regarded as a factor that is produced by the investment of human capital, then the construction of an aggregate is just as difficult as in the case of capital. Whether the aggregation of different types of labour is thwarted by the same difficulties as those encountered in the aggregation of capital depends therefore on the question whether labour is an original or a produced factor of production.[36]

We must therefore conclude that, like the microeconomic production function, the macroeconomic production function fails as a condensed representation of the efficient technical possibilities because of the economic heterogeneity of the production techniques. This conclusion is confirmed when, as well as the state of technology, technical development is taken into account. New techniques and new products enlarge the set of technical possibilities, so that the construction of an aggregated production function describing the development in the course of time is equally impossible.[37] Ferguson's opinion, noted in Section 8.11, that technical development can be expressed by a macroeconomic production function that describes the gradual course of the technique in time, stems from his mistaking a temporal continuity for a causal connection. The expansion of the technical possibilities in the macroeconomic sense cannot be easily derived from the situation in the previous period, since the technical possibilities involved are entirely new. Even though the effect on for example productivity may be gradual, the identification of technical development with a macroeconomic production function shifting in time betrays a rather naive concept of technical development.

A better idea of technical development on the macroeconomic level can be attained by evoking axiomatic production theory once more. A macroeconomic set of technical possibilities can in principle be constructed from the sets for all the firms by simply adding up these sets. The problems arising here are comparable to those discussed by Debreu when putting price theory on an axiomatic basis. Physically identical products and means of production cannot be simply aggregated over the firms, since the economic functions involved may differ according to time and place, nor does a closed production set for each firm guarantee a closed overall production set.[38] The sum of the efficient combinations for the individual firms need not therefore give efficient combinations for the whole branch of industry or the whole economy. If technical development is now introduced, it is even less likely that aggregation, which need not give efficient combinations at a given state of technology, will lead to a closed overall production set.

9.3 CHOICE OF TECHNIQUES

Some aspects of the macroeconomic theory of the entrepreneurs' choice from the available technical possibilities can be further elucidated with the aid of wage-interest curves and the wage-interest frontier. All the entrepreneurs are assumed to be able to make a consumer good and a capital good by two techniques α and β, so that two wage-interest curves corresponding to these techniques can be constructed, using the assumptions mentioned before. This produces three possibilities, which are discussed below in turn.[39]

First of all, it is possible that technique β completely dominates technique α. This means that the whole of the wage-interest curve for β lies to the right of that for α, so that the curves do not intersect in the positive quadrant. Technique β is then used at every interest rate r, since it always gives a higher real wage rate. The wage-interest frontier coincides here with the wage-interest curve β. If technical development provides a new technique γ that is superior to technique β, then the wage-interest frontier is displaced to the right. If we assume perfect competition and sufficiently fast adjustment, technique γ will in fact be used in practice. This situation, in which the latest technique is always used, was the basis for our analysis of Marx's concept of technical development in Chapter 3. Neutrality of technical development in Harrod's sense, which was discussed there, means that the wage-interest frontier $F(w,r) = 1$ is changed into $F(e^{-kt},w,r) = 1$.[40]

The second possibility is that the wage-interest curves α and β intersect each other once in the positive quadrant at P. If the interest rate corresponding to this point is denoted by r', the wage-interest frontier is formed by that part of the wage-interest curve β for which values of r are smaller than r', and by that part of the wage-interest curve α for which values of r are greater than r'. The choice of technique then depends on the value of the rate of interest: if r is lower than r', the entrepreneurs choose technique β, and if r is higher than r', they choose technique α. If r equals r', we have the special situation, characteristic of the point P, in which both techniques α and β give the same maximum value of the real wage, so that the entrepreneurs' choice of technique is not determined. The introduction of a new technique γ whose wage-interest curve shares one point with the original wage-interest frontier leads to a new wage-interest frontier, one section of which coincides with the original one. While in the case of neutral technical development the relevant wage-interest curves can be assumed to intersect each other at one point at most, more than one point of intersection can easily arise when a non-neutral technical development is assumed.

When the wage-interest curves α and β intersect at two points the situation shown in Figure 6 (see page 152) is obtained. Here we have two switch-points, and the wage-interest frontier indicates the return of techniques. A technique that has been chosen by the entrepreneurs at relatively low values of interest becomes eligible for reintroduction at relatively high values of r, while for intermediate values of r a different technique is chosen. At the two switch-points the entrepreneurs' choice of technique is again indeterminate.

The third possibility is that an infinite number of techniques are available; in this case, each point of the wage-interest frontier is a point of contact with the wage-interest curve, so that there is another profit-maximizing technique belonging to each value of the rate of interest r. A change in the latter therefore means that the entrepreneurs will employ a different technique on the assumptions mentioned before.

In the macroeconomic equilibrium situation outlined above the choice of technique depends to some extent on the interest rate, without the interest rate acting as a yardstick of the relative scarcity of capital. It is very important to realize that the rate of interest cannot be obtained endogenously from the description of the technical possibilities; but even if it is given exogenously, it is still not possible to draw easy conclusions about the technique that the entrepreneurs will choose. If the given value of the rate of interest corresponds to a switch-point, certain assumptions are needed about the preferences of the entrepreneurs in order to decide which of the equally profitable techniques they will choose. These preferences also play a role in so far as in all the other cases a change in the given rate of interest would cause a change-over to another technique. The preferences then relate to the decision about the immediate complete depreciation of the existing capital goods and the possibilities of manufacturing new capital goods. Besides these complications connected with the transition to a new technique, there is the problem of the degree of utilization of labour, which varies from one technique to the next. These last considerations arise in particular when each point on the wage-interest frontier represents a different technique.

Furthermore, the determination of the choice of technique from the choice of the rate of interest is somewhat arbitrary. There is a relationship between the per-capita consumption and the rate of growth of capital g, which parallels the relationship between the real wage and the interest rate. It is therefore also possible to regard the rate of growth of capital as an exogenous quantity, or to determine it in such a way that the per-capita consumption is maximized. If the latter is done, the criterion for the choice of technique is the optimum allocation over time, and the welfare economic background to this will be discussed in Chapter 12.

The different assumptions that can be made about the determination of the rate of interest (or the rate of growth of capital), real wage (or per-capita consumption) and population growth illustrate that the macroeconomic description of the technical possibilities with the aid of the wage-interest curves does not yield a closed system. Since we are mainly involved here with the problem of expressing the technical aspects of the economic process, utility and demand functions should be introduced in order to complete the open system in principle and to determine the equilibrium values of the prices of goods.

In Sraffa's approach, this equilibrium mechanism is replaced by the hypotheses that the means of production are reproduced and that the distribution of the surplus resulting from production is determined exogenously.[41] Here, the choice of technique in which the real wage is maximized at a given rate of interest r depends not on the assumption of perfect competition, but on the condition that a production method is used whereby at least the means of production used up in the production process are reproduced. Sraffa's theory therefore agrees with the Classical idea of a cyclic process in which goods are produced continuously with the aid of goods, and it disagrees with 'the view presented by modern theory, of a one-way avenue that leads from "factors of production" to "consumption goods".'[42]

Without taking up a definite position in this fundamental controversy about the bases of theoretical economics, one should notice that the discussion has so far been about choosing a technique from a given set of technical possibilities. If a process of non-neutral technical development is now introduced into the argument, then two results can be expected: current technical possibilities will be dominated by new ones, and the construction of additional wage-interest curves gives new switch-points, at which different techniques are equal contenders for application at a given rate of interest. In these cases, finding the optimum allocation of the means of production requires further criteria in Sraffa's treatment, as it does in the neo-classical version of the theory. Since it is difficult to obtain these criteria from economic categories other than the preferences of individuals, the following analysis will start from the assumption that the demand for goods plays an independent role in the economic process.[43] This analysis of the causes of technical development is likely to be more fruitful both in a macroeconomic context, in which the hypothesis of perfect competition is mostly retained, and in a microeconomic one, in which other forms of competition are also taken into account.

The social significance of the Cambridge controversy can be understood better if it is realized that the West has long been practising neo-classical economics, in which income distribution depends on production performance. Since the 1960s, however, there has been an increasing tendency to regard income distribution as an independent social entity, reflecting social bargaining rather than productivity. This belief, associated with Cambridge (England), has influenced the composition of the factors of production in the economic process, and more and more labour is being displaced because of high real wages. If western economies want to return to a state of more or less full employment, the real wage rate must again be geared to productivity.

9.4 NEUTRAL TECHNICAL CHANGE

Although serious objections can be raised against the aggregate production function on the grounds of the heterogeneous nature of the production methods, this representation of the technical possibilities at the macroeconomic level is nonetheless widely current in economic theory. A macroeconomic production function is constructed that corresponds to the microeconomic relation between production and the means of production, without accurately specifying how this can be derived from the set of heterogeneous production processes.[44] This approach is not entirely satisfactory from a theoretical point of view, but so much empirical research has been carried out with the aid of this device that we cannot ignore it. In empirical work one goes further than is necessary for a description of the state of technical possibilities, and expresses technical change on a macroeconomic level by introducing time into a 'well-behaved' neo-classical production function, along with capital and labour, thus obtaining $F(K,L,t)$. Technical change is then manifested in a higher efficiency of the factors of production capital and labour.[45] In this treatment, a shift in the production function is neither neutral nor non-neutral. But since the shift in the production function in time is identified in the relevant literature with technical change, these possibilities are usually called neutral and non-neutral technical change.

The discussion about the neutrality of a technical invention has long been dominated by the question of the influence of mechanization on capital and labour in the production process, and on the distribution of output between capitalists and employees. This takes us back to Ricardo, who in our opinion initially viewed the introduction of machines as neutral, since he believed that the employees would benefit from the advantages of mechanization to the same extent as would other sections of the society. The gist of his analysis in the third edition of the *Principles*, however, is that capitalists and workers are not affected by mechanization in the same way, for the real profits increase while the wage fund decreases;[46] technical change is then non-neutral.

Wicksell, who—like Ricardo—did not use the terms neutral and non-neutral, took Ricardo's views further by expressing the effect of technical change on income distribution with the aid of the theory of marginal productivity.[47] He clearly states that an increase in total output as a result of technical change need not mean that the marginal productivities of labour and capital increase to the same extent.[48] It is interesting that Wicksell criticizes Ricardo for overlooking the fact that, owing to a fall in wages caused by the dismissal of workers, the old method of production again becomes after a time the more profitable one, so that an equilibrium is reached in which the marginal productivity of labour in the old and new production methods are the same, and a single level of wages is established.

As we have seen, however, Ricardo assumes fixed proportions in this case, so that even the drop in wages does not represent a sufficient condition for an equilibrium at the level of full employment (see Section 2.5). Looked at like this, Ricardo's opinion on the neutrality of technical change is more compatible with Harrod's interpretation than with Hicks'.

Nevertheless, how the marginal products of labour and capital are influenced by the introduction of a new method of production is very important for Wicksell, since the income distribution between the owners of the factors of production can be derived from it. Exactly the same view is held by Pigou,[49] who introduced the terms neutral, capital-saving and labour-saving inventions. Capital-saving inventions raise the marginal product of labour more than they raise the marginal product of capital (while labour-saving inventions do the opposite), and the availability of an extra unit of capital has a greater effect on production than does the availability of an extra unit of labour. The level of output has become as it were qualitatively less dependent on labour. Hicks used Pigou's terms in his early work on the theory of wages,[50] and ever since economists have spoken of neutral technical change in Hicks' sense whenever this development leaves the ratio between the marginal products of labour and capital unchanged. Thus for the production function $F(K, L, t)$, $F(K, L, t) = A(t)$. $F(K, L)$; in other words, the production function shifts in each period.

Harrod gave his first description of neutral technical change in a review of a book by Joan Robinson,[51] in which she used Hicks' classification. Harrod argues that neutrality in Hicks' sense is difficult to demonstrate because of the thorny question of the measurement of capital, and proposes that one speak of a neutral technical change if the invention does not affect the production period at a given rate of interest. In her response,[52] in which she devotes only a footnote to the measurability of capital, Robinson shows that the neutrality of an invention in Harrod's sense has the same effect as an increase in the quantity of labour at a constant technology, in other words the new technique increases the productivity of labour. This is what is called in everyday language labour-saving technical change. In later publications, Harrod further clarified his concepts and established their connection with his growth model.[53] A neutral technical change leaves the production period and the capital coefficient unchanged at the same rate of interest; the background to this definition emerges from the following quotation: 'As I have chosen to approach the dynamic problem by asking what rate of increase of capital would be consistent with certain rates of increase in other parts of the system, it has seemed simplest to define a neutral stream of inventions as one which shall require a rate of increase of capital equal to the rate of increase of income engendered by it'.[54]

It is not absolutely clear whether Harrod is thinking here of *ex ante* relations between production and the factors of production, or of the equilibrium paths of national income. The first interpretation is often related in the literature to the idea that Harrod starts with fixed proportions and constant returns to scale, so that the capital coefficient is constant. Neutral technical change in Harrod's sense is then shown in a new value of the labour coefficient. However, Uzawa[55] has shown that a neutral technical change in Harrod's sense is generally involved whenever the production function $F(K,L,t)$ assumes the form $G(K,A(t),L)$. This means that neutral technical change in Harrod's sense can be put on a par with an increase in the working population. Uzawa has also shown that the Cobb-Douglas production function $A(t).K^{\beta}.L^{1-\beta}$ is the only one in which technical change is neutral both in Harrod's sense and in Hicks'.[56]

Our discussion of neutral technical change has so far been concerned with the whole production function, so that each point of the new production function for which the criterion of neutrality applies corresponds to a point of the original one. However, there is also another possibility, namely that the definitions given refer only to certain points of the production function. The obvious thing to do in this case is to concentrate on the equilibrium situations. The discussion is then no longer about the neutrality of inventions, but about the neutral character of the applications, since each equilibrium point signifies the optimum choice from among the technical possibilities.[57] The characteristics of an equilibrium are determined not only by the production function used, but also by the entire set of relations—the model—that serves to characterize the situation. It is understandable that, to some economists, Hicks' and Harrod's definitions of neutral technical change boil down to the same thing,[58] since in some models a technical change can be neutral both in Harrod's sense and in Hicks'. This means that the comparison of two equilibrium situations in the framework of such models leads to the conclusion that an actual technical change takes place which is neutral according to both these authors. This statement then refers to a comparison between an equilibrium position on the old production function and an equilibrium position on the new one, and therefore it does not generally apply to other combinations of the factors of production. It is not surprising therefore that in other models the two types of neutrality do not coincide.[59]. Birg[60] wrongly suggests that for this reason the description of neutral technical change should be independent of a particular model, since one must *always* start with one or more magnitudes that are treated as constant.

We should also examine against this background the other interpretation which can be given to the quotation from Harrod's work, which reveals a different idea of neutral technical development. According to this interpretation, Harrod has in mind the derivation of the equilibrium conditions for a process of balanced growth, and his model is readily compatible with a neo-classical production function. His definition of neutral technical change is formulated so that technical change does not disturb a process in which output, capital and labour increase to the same extent. This interpretation of Harrod's ideas has been given mainly by Helmstädter,[61] who believes that what is involved here is the optimum capital coefficient at balanced growth, the value of which remains the same in the case of neutral technical change. Starting with an intertemporal production function, we illustrate this interpretation of Harrod's view in Figure 7 with the aid of a diagram used by Wicksell (see Section 5.5).

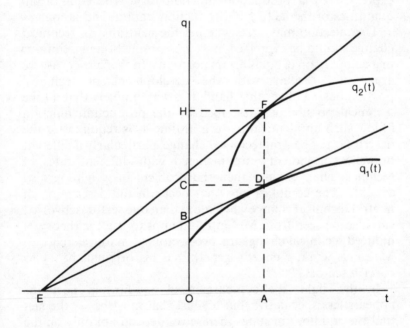

Figure 7

Points D and F denote the original and the new equilibrium point respectively, the rate of interest is given by OE and the optimum production period by OA. Thus for points D and F the optimum production period remains the same at a constant rate of interest. Viewed thus, Harrod's description of neutral technical change is closely connected with intertemporal production theory. Harrod's definition can also be illustrated by a comparison of equilibrium situations with the aid of non-intertemporal production functions.[62]

Neutrality in Harrod's sense is mirrored by Solow's concept,[63] according to which the production function $F(K,L,t)$ shifts in time to become $G(A(t),K,L,)$. We are dealing here with a capital-augmenting technical change, which is usually described by the somewhat vague term 'capital saving'. This type of technical change is important mainly when advances in technology are embodied in new capital goods. Neutrality in Solow's sense can of course refer either to all the technical possibilities that are expressed by a production function or to the equilibrium quantities of the factors of production, capital and labour.

The question may arise whether the neutrality of technical change should be expressed in *ex ante* production functions or in a comparison of equilibrium positions. In the former case we are really dealing with the development of technical possibilities. On the other hand, it is very unlikely that all the technical possibilities described by the production function follow such an identical pattern in time as is required for the description of neutral technical change, particularly if different branches of industry are included with different rates of technical change, since then the problem of weighting also arises.[64] The complications that occur in the description of neutral technical change when more than one sector is involved can also be seen from McCain's attempt to use the theory of induced technical change for two sectors.[65] With reference to McCain's work, Kennedy generalized his original theory to cover n sectors.[66]

In this light, the comparison of equilibrium positions appears less speculative, but against that must be set the fact that the equilibrium states themselves depend not only on the technical possibilities, but also on economic factors and on the behavioural hypothesis introduced. In this case the neutrality of technical change implies not only a certain evolution of the technical possibilities but also a particular behaviour on the part of individuals. This viewpoint is not without significance for a satisfactory assessment of empirical research since this is concerned with monitoring the development of equilibrium situations over the course of time.

Our conclusion is that each of these two methods has its own relevance, provided that they are not confused. It should be added, however, that the second method is the appropriate one for the economic analysis of the actual growth process. It is interesting therefore that Beckmann and Sato,[67] who have done some important empirical investigations of the various types of technical development, arrive at an extension of the cases of neutrality on the basis of a number of neo-classical assumptions. They obtain generalizations of neutrality in the sense of Hicks, Harrod and Solow by assuming that technical change leaves the elasticity of substitution intact at a constant ratio between labour and capital and between capital and output and at a constant distribution of incomes between social groups. Drawing on this work of Beckmann and Sato, Brubaker has recently introduced the concept of multi-neutrality of technical change, where the latter exhibits neutrality in the sense of Hicks, Harrod and/or Solow.[68] Nôno has developed a still wider generalization with the aid of Lie algebra.[69]

The discussion of some criteria of neutral technical change shows that the objections to the use of an aggregated production function apply equally to the description of technical change with the aid of a shifting aggregated production function. Furthermore, the actual course of technical development in the various sectors of an economy is so diverse that it must be regarded as an unusual coincidence if the aggregated production function—provided that it exists at all—reflects a neutral technical change in the sense of Hicks, Harrod or Solow.[70]

9.5 KALDOR'S ALTERNATIVE

The difficulties encountered in the description of the available technical possibilities with the aid of a production function led Kaldor to develop an alternative theory. In his opinion, a distinction between a movement along a production function and a shift in it is arbitrary and artificial.[71] The use of more capital per employee is inevitably associated with the introduction of better techniques, and—conversely—technical innovations generally call for the use of more capital. Instead of explaining the growth of production by the accumulation of capital and the advances of technology, Kaldor introduces a relationship between the growth of production and the growth of capital per head, whieh he calls the 'technical progress function' and which increases monotonically, though at an ever decreasing rate.

The way Kaldor introduces this function can give the impression that this is an empirical law. This emerges from Krelle's interpretation,[72] according to which only one point on the production function is realized in practice in any one period, and Kaldor's function is a concatenation of these equilibrium points in a time sequence. This interpretation holds that the technical progress function reflects the actually chosen technical possibilities, which are in turn again dependent on the increase of technical knowledge, the formation of capital, and relative prices. In this form, the technical progress function is difficult to compare with a shifting *ex ante* production function. With a linear technical progress function, which can be reduced to a Cobb-Douglas production function,[73] the generated production function must deal with equilibrium situations and therefore be considered an *ex post* relationship.

In contrast to this *ex post* interpretation of Kaldor's function we should note that the technical progress function is a component of a larger model aimed at deriving the rate of growth of production.[74] For this reason it seems more sensible to treat the technical progress function in an *ex ante* sense.[75] When this is done, the linear variant of Kaldor's function is compatible with a shifting *ex ante* Cobb-Douglas production function, so that the previous objections can be repeated as regards the portrayal of technical change. In the non-linear case, the combining of technical change and the accumulation of capital is open to strong objection, since these phenomena, which—as Kaldor rightly remarks—are often strongly connected in practice, should be kept apart in economic analysis, which aims to study their effects on the growth of production separately.[76] Insofar as technical development is embodied in new capital goods, this can be expressed more explicitly by making a further distinction amongst the capital goods.

Our conclusion is that Kaldor's technical progress function is not an improvement on a shifting production function, because—as well as being open to the same objections as the production function—it introduces a complication by not separating technical change from the accumulation of capital.[77]

9.6 CONCLUSIONS

The neutrality of technical change in the sense of Hicks, Harrod or Solow applies both to the shift in the *ex ante* production function and to the comparison of equilibrium positions in the course of time. In this second case, the equilibrium quantities of output and the factors of production depend in particular on the values of the factor prices and on the hypothesis concerning the behaviour of producers. The macroeconomic interpretation of this situation is less simple than the microeconomic one, since—apart from the question of whether a macroeconomic production function exists at all—one should start with such an aggregate enterprise behaviour that an equilibrium point is formed, as in the microeconomic derivation.

In the microeconomic case, one might try to describe technical change directly in terms of its effect on the quantities of the factors of production in the equilibrium position, and this is what Salter has done.[78] He describes a neutral technical change with the aid of the ratio between labour and capital at equilibrium before and after the technical change at constant factor prices and on the assumption that the total cost is minimized. This approach also makes it possible to define the concepts of labour-saving and capital-saving change directly by simply comparing the required equilibrium quantities of the factors of production.

Salter's descriptions are particularly important for empirical research on technical change, since they fit in well with the actual state of affairs, but they are less suitable for theoretical analysis because of the unavoidable intertwining of technical and economic circumstances, inherent in the *ex post* interpretation. If Salter's description is to have a macroeconomic meaning, it must be derived from the microeconomic concept, in connection with which there are some considerable weighting problems. However, the existence of a macroeconomic relation or concept can also be postulated by analogy with the corresponding microeconomic relation or description.

Salter's view that substitution and technical change are difficult to separate in practice is associated with the framework of his work, dictated mainly by the requirements of empirical research. In this respect, Salter comes close to Kaldor, although Salter does not reject the concept of the production function. Ott[79] has proposed a classification of technical change that seems at first sight to be closely related to that of Salter's. Ott starts with the identity $V \equiv qK + l.A$, where V is the total cost, and q and l are the prices of the factors of production, capital (K) and labour (A). Since Ott does not introduce a production function and thus does not derive any equilibrium positions, we are dealing here with completely arbitrary changes in K and A, so that his attempt to separate technical change from substitution by keeping q and l constant is frustrated.[80] On closer inspection, therefore, there is no real similarity to Salter's work, since in his theoretical framework Salter does manage to separate substitution from technical change.

We find again and again that attempts to construct macroeconomic quantities and relations from microeconomic phenomena meet with great theoretical difficulties. An added complication is that it is logically inadmissible to verify the postulated *ex ante* relations on the basis of empirical observations. The comparison of equilibrium positions always involves the danger that the empirical relations between the quantities are identified with the corresponding *ex ante* connections, which only form a part of the complex of relations from which the *ex post* situation results. Technical development in the sense of an expansion of the technical possibilities is such a diverse process from the microeconomic point of view that— purely theoretically—it can be described more or less adequately only by assuming that new production functions arise all the time. It follows from this that, on the macroeconomic level, empirical research on technical development is left hanging in the air, since it is based not only on the existence of a macroeconomic production function, but also on its shifting in the course of time. Furthermore, the analysis takes on a speculative character because it mostly starts with the assumed neutrality of technical change. A tension is therefore set up between the results of the macroeconomic empirical investigations, which only give an overall impression of the actual effect of using new techniques in terms of a few macroeconomic quantities, and the continuously changing and expanding microeconomic set of technical possibilities which embodies detailed technical knowledge of both a qualitative and a quantitative character. Against this background there have been various discussions of the question whether a production function shifting in time can be used to characterize technical change, the views of Solow and Robinson being the clearest statements of the opposite sides of this argument[81] (i.e. of the macro- and the microeconomic sides).

We believe that this tension is resolved only if the macroeconomic approach is removed from economic analysis and all economic phenomena are explained on the basis of the behaviour of individuals. For the time being, empirical macroeconomic research should be accepted as a temporary bridge between the theoretical macroeconomic concepts which are still lacking and that part of the theory which can be verified empirically. Macroeconomic relationships and concepts should preferably be based on microeconomic analysis in a satisfactory manner, but from the point of view of the development of economics we should not abandon the macroeconomic approach if it cannot be based on microeconomic analysis. The portrayal of technical change by a shifting macroeconomic production function exemplifies such an *ad hoc* macroeconomic approach, which has nevertheless enriched our understanding of reality.

Mention should be made of Houthakker's attempt—which went unnoticed for some time—to derive the macroeconomic production function from the microeconomic *ex post* distribution of the (variable) means of production per unit end product.[82] Solow[83] stressed the great importance of Houthakker's analysis for production theory, Levhari[84] and Sato[85] have generalized Houthakker's results, and Johansen[86] has recently published an important book, in which, in the footsteps of Houthakker, the connection between *ex ante* and *ex post* micro- and macroeconomic production functions is demonstrated on an impressive scale. According to Johansen, the gist of Houthakker's view is that a macroeconomic production function capable of accommodating technical change can be derived from the actual capacity distribution, which is based on the chosen combination of the means of production in the course of time. This gives a sounder foundation to econometric research on dynamic macroeconomic production functions in particular than does the approach in which the macroeconomic relations are constructed exclusively by analogy with microeconomic ones. While this solves the aggregation problem, it has the great disadvantage, in the static approach, that the set of technical possibilities on the macroeconomic plane consists of microeconomic equilibrium solutions which depend partly on the *ex ante* microeconomic production function and partly on the equilibrating mechanism involved. However, this approach completely fits in with the view, developed in Section 8.7, that in a dynamic analysis one cannot avoid explaining the current production function by economic decisions taken in the past. One aspect of such an analysis is the interaction of technical change and capital accumulation.[87]

A more general conclusion to be drawn from this chapter is that the Cambridge approach to economics has reinforced the view that technology in the static sense and technical change in the dynamic sense are essentially endogenous phenomena.[88]

10 Technical change and economic growth

The fun in scientific research, as in life itself, is in the dynamics and not the statics of the problem.

Samuelson[1]

10.1 INTRODUCTION

We shall now examine various widely discussed types of technical change which have so far been mentioned only in passing. The framework for this discussion is provided by the production function, and technical change is viewed as a shift in this function. Growth theory is introduced by examining some hypotheses about the growth of the working population and capital formation alongside the assumptions about the nature of technical development.

It has already been mentioned, in connection with Kaldor's models, that technical change embodied in capital goods can be directly portrayed by the production function.[2] Using a model constructed by Solow, we shall show in Section 10.2 that in the case of technical change embodied in capital goods, the level of production in period t depends on the period, v, in which the capital goods were installed.[3] If technical change is not embodied in capital goods, on the other hand, then the level of production in period t depends—as far as technical change is concerned—on the value of the general trend factor in period t. Both these cases can be combined with the neutrality of technical change in the sense of Hicks, Harrod or Solow. For the case of disembodied technical change, this has already been shown in Section 9.4.

The case of technical change embodied in capital goods can now be derived by replacing period t by period v. If the production function is $Q(v,t) = A(t).F[K(v,t), L(v,t)]$, technical change is neutral in Hicks' sense and is of the disembodied type. In the case of embodied technical change that is neutral in Hicks' sense, the production function is $Q(v,t) = A(v).F[K(v,t), L(v,t)]$.

Another distinction is that between exogenous (or

autonomous) and endogenous (or induced) technical change. In the case of the former, technical change as such is not explained, and the rise in the level of production is simply attributed to the passage of time. In the case of endogenous, or induced, technical change, the expansion of technical possibilities is explained explicitly by one or more economic factors, such as (a) long-term changes in the ratio between the prices of the factors of production, (b) learning processes concerning production, and (c) investment in education and research. These will all be discussed later.

It has been stressed before (Section 8.11) that technical development is partly determined by economic factors, so that it is not one of the external data of economics. For this reason, endogeneity is a characteristic of technical development itself. In the literature under discussion, on the other hand, the term 'endogenous' refers mainly to the way technology is viewed. Although this reflects the actual advances of technology up to a certain point, it should be remembered that treating technical development as an exogenous entity in the framework of a model does not mean that it is independent of economic factors in reality.

10.2 TECHNICAL CHANGE EMBODIED IN CAPITAL GOODS

Solow[4] labels capital goods by the year of their production ('vintage') and denotes the amounts of capital goods from year v that are still in use in year t as $K(v,t)$. The amount of labour combined in year t with capital goods from year v is denoted by $L(v,t)$. The capital goods from year v embody the state of technology at that period, and technical change is not manifested in an increase in the productivity of these capital goods. Using a Cobb-Douglas production function for year v, we obtain:

$$Q(v,t) = B.e^{\lambda v}.L(v,t)^{\alpha}.K(v,t)^{1-\alpha}$$

The rate of embodied technical change is expressed by $e^{\lambda v}$, and the earlier the capital good was made, the lower its productivity. This shows that the productivity of capital goods depends not on the current date (year t), but on the year in which they were made (year v). In the case of the Cobb-Douglas production function, embodied technical change is neutral in Solow's and Harrod's sense.[5] The disembodied component of technical change can be easily introduced by incorporating a trend factor $e^{\lambda v + rt}$, where r is a measure of the rate of disembodied technical change. It is also possible to construct a model in which technical change is embodied both in capital goods and in labour, with education playing an important role.[6]

In Solow's model, restricted to capital goods, the total production period t is the sum of all the quantities produced in years v, it being assumed that there is a homogenous end-product. In principle, we would have to carry out the summation between now and infinity in the past. If the vintage v is regarded as a continuous variable, the total production in period t, i.e. $Q(t)$, is given by the integral:

$$Q(t) = \int_{-\infty}^{t} Q(v,t).dv$$

The total amount of labour needed in period t, i.e. $L(t)$, can be similarly written as an integral, that is a sum of the labour inputs used up in all the vintage years:

$$L(t) = \int_{-\infty}^{t} L(v,t).dv$$

It is assumed here that the total demand for labour is always equal to the total supply, $L_0.e^{nt}$.

An interesting feature of Solow's model is that it permits an expression to be derived for the total stock of capital goods $K(t)$, and this expression in turn enables us to construct, in the present case, an aggregate macroeconomic production function. Solow assumes for this purpose that there is only one wage rate $w(t)$ in each period. In each period, the available labour is now distributed over the various vintages of capital in such a way that the marginal product of homogeneous labour is the same for each year of vintage.[7] On the basis of this equilibrium mechanism, the total production $Q(t)$ can be written as a Cobb-Douglas production function, where $L(t)$ and

$J(t)$ denote total labour and 'capital' respectively. This measure of capital is provided by the integral:

$$J(t) = \int_{-\infty}^{t} e^{\sigma v}.I(v).dv$$

where $\sigma = \dfrac{\lambda}{1-\alpha} + \delta$, $I(v)$ is the total investment in the year of vintage v, and δ is the rate of depreciation (per cent).

In Solow's treatment, the wage rate $w(t)$ is determined in a labour market of perfect competition. Since labour is homogeneous a market supply function can be constructed for it, the counterpart of a market demand function for labour. Is too much attention given here to the technical homogeneity of labour and is the market supply curve not construed in too mechanical a way? Since labour is combined with capital goods made in different years in the past, it seems necessary to take into account the various subjective preferences the suppliers of labour have for old or new mechanical equipment. These different preferences make the labour supply economically heterogeneous, so that as many labour supply curves can be constructed as there are years of vintage. If these subjective preferences play an important role in the market, we are faced with a wage structure rather than a single wage rate.[8] This undermines the basis for the derivation of the total labour demand function from the macroeconomic production function, which rests on the hypothesis of a single wage rate in each period. An objection to Solow's aggregation is therefore that not enough attention is paid to the effect a differentiation of capital goods has on labour.[9] Caution is therefore necessary when working with growth models based on Solow's analysis.[10]

The objection raised against the method of aggregation does not alter the fact that this approach gives one an insight into the way in which embodied technical change can be expressed.[11] The main points are that capital goods which have been installed do not experience a rise in productivity, and that technical change is embodied in new capital goods. A production function, constructed for each vintage, shows the relationship between the production period t, labour, and the capital goods installed in period v. The question now is how well this production function portrays the technical possibilities, since the amount of capital goods actually installed in period v is based on a choice from the possibilities existing at that time. In period t, however, there is no longer a choice as regards capital goods, since all that remains to be decided is how much labour the remaining capital goods will be combined with. From this point of view, the production function used by Solow, $Q(v,t) = B.e^{\lambda v}.L(v,t)^{\alpha}.K(v,t)^{1-\alpha}$, is an *ex post* production function which portrays only limited technical possibilities, since the capital from year v is already actually present. The neutrality of embodied technical change in Solow's and Harrod's sense then concerns only the equilibrium quantities of the capital goods. The production function $Q(v,v)$ $= B.e^{\lambda v}.L(v,v)^{\alpha}.K(v,v)^{1-\alpha}$ can then be regarded as an *ex ante* production function, since it is now specified which combinations of labour and capital in period v yield a certain amount of product in the same period v. This *ex ante* interpretation is ruled out if in the next period the production Q in vintage year v is explained by the labour and capital inputs in period v. It is hardly surprising therefore that it was the recognition of embodied technical change in the theory of growth that urged economists to construct models in which a certain degree of substitution is assumed *ex ante*, and fixed proportions are postulated *ex post*.[12]

When technical change embodied in capital goods plays an important role in a growth model, we no longer have an explanation of the expansion of technical possibilities, but a description of the effects of the application of technology on the growth of production. The recognition of embodied technical change implies that a choice has been made from the available technical possibilities.

10.3 ENDEOGENEOUS TECHNICAL CHANGE
Technical change embodied in capital goods is exogeneous, because the increase in the productivity of capital goods is here attributed solely to time. The reasons for the higher productivity of the recently installed capital goods are not analysed, and decision-making is not involved. However, this approach is less non-committal than the one in which technical change is *not* embodied in capital goods. In this latter case the

passage of time automatically ensures that the various combinations of the factors of production give a higher output, whereas in the case of embodied technical change the construction of new capital goods is at least necessary.

It will already be clear from this that the distinction between exogeneous and endogeneous technical change refers not so much to the factual aspects of technical development as to the depth of the analysis of the phenomenon. In the simplest case, technical change is regarded simply as a shift in the production function in the course of time, but a closer examination of the factors influencing this shift leads to a more refined view of technical change. For a thorough understanding of the debate about endogeneous technical change, one must examine whether the aim is to explain the expansion of technical possibilities or to analyse the entrepreneurs' choice of a particular type of technical development. This second case will be examined first, although it can be argued that we are not dealing here with a proper endogeneous technical change, since the expansion of technical possibilities is not explained but assumed to be given. Here we are typically faced with the case in which the nature of technical change is linked with a change in the price of the factors of production. In this generalized form, the problem is not a new one and is reminiscent of Ricardo's famous statement about permanent competition between labour and mechanization. Hicks has pointed out that a change in the relative prices of the factors of production can act as an incentive for inventions 'directed to economizing the use of a factor which has become relatively expensive'.[13] Here the distinction between inventions and their applications fades, since the former are made with a view to the latter, even if it is true that an invention triggered off by a rise in wages is not utilized until much later, if at all, for reasons lying outside economics. Hicks himself believes that, in Europe in the last two centuries, induced labour-saving inventions have been more numerous than induced capital-saving ones.[14]. Salter disputed this view, saying that the aim of the entrepreneur is to minimize his total cost, and it is immaterial to him whether the increase in the wage bill can be counteracted by a labour-saving method or by a capital-saving one; there is no reason to concentrate on labour-saving techniques, unless they are easier to acquire than capital-saving ones.[15] In contrast, Fellner believes that the former are more profitable in the long-run than the latter because of expected shifts in the relative scarcity of labour and capital.[16] It is interesting to note that Fellner uses the term 'innovation' instead of 'invention' in the title of his study.

Kennedy, who was the first to publish a mathematical and economic treatment of this topic, shifts the emphasis entirely from inventions to applications, and by doing so moves a long way from Hicks' original analysis.[17] Kennedy assumes a technical change in which the amount of at least one of the factors of production decreases in percentage terms. If there are two factors of production and if the changes in their amounts in percentage terms are denoted by p and q, then the possible innovations are represented by $Q(p,q) = 0$. Plotting p against q in Cartesian coordinates gives an innovation curve which is convex with respect to the origin. If the cost of labour and capital are introduced, it is possible to choose the least expensive innovation.

The main point here is that this curve showing the technical possibilities forms the basis of the analysis and is not derived from changes in the prices of the factors of production. We are no longer dealing with induced or endogeneous technical change, but with the choice of the best technique of production in a macroeconomic setting. It may be asked how Kennedy's analysis differs from the usual determination of equilibrium with the aid of isoquants in the theory of production. The question is all the more pertinent as Kennedy did not introduce a production function, hoping that 'the innovation-possibility frontier might be able, so to speak, to swallow up the traditional production function and replace it altogether'.[18] Kennedy's approach blurs the distinction between the choice of a certain method of production at a given state of technology and the choice of a certain type of technical change from all the possible ones.

The studies that followed Kennedy's work elucidated this last point, but did not deal with the objection that technical change is not derived here from changes in the prices of the factors of production, but constitutes instead a starting point in the formal analysis.[19] Using a production function with capital-augmenting and labour-augmenting technical change, i.e. $Q(t) = F[A(t)K, B(t)L]$, one can introduce more specific assumptions concerning the relationship between the possible capital-saving and labour-saving technical changes, i.e. between $A(t)$ and $B(t)$. This technical possibility frontier, $Q[A(t), B(t)] = 0$, is concave and bounded. At any point, there is a choice of the same alternative combination of labour-augmenting and capital augmenting technical change in the macroeconomic context. The decision to accelerate one type of technical change implies a slowing-down of the other. To find the optimal combination of the two types of technical change, an analytical apparatus is available, provided the necessary assumptions are made about the cost of labour and capital. In this way, the production function, which, as such, describes the state of technical possibilities, is clearly distinguished from technical change which is manifested in a decrease in the inputs of the factors of production, reflected in $A(t)$ and $B(t)$. What we have here is not the optimal choice of a certain isoquant of the production function, but the optimum rate of change, $[A^*(t), B^*(t)]$.[20]

We have so far discussed the choice between capital-saving and labour-saving technical development, but technical development leads not only to a possible saving of the factors of production, but also to the emergence of new products. It is understandable therefore that some attempts have been made to extend the Kennedy–Weizsäcker model of induced innovation and growth to include new-product innovation. Thus, inspired by Lancaster's[21] new approach to consumer theory, McCain[22] has tried to introduce product innovation into a Kennedy–Weizsäcker model.

Various types of optimal technical change can be derived, according to the shape of Kennedy's technical possibility frontier. In the case of a linear Kennedy frontier, technical change is neutral in Harrod's sense, provided that a 'well-behaved' neo-classical production function is used.[23] Conversely, the attributes of the technical change determine the shape of the curve for the technical possibilities, as is clearly demonstrated in Chang's study.[24]

The above analysis hardly deals with the connection which, according to Hicks, exists between a steady increase in the wage bill and the technical change triggered off by it. The theory is therefore too formal, and treats technical development more as an entity that explains things than as one that itself requires an explanation. Yet these theoretical studies have stimulated some new lines of study which place more emphasis on the factors that determine the creation of new technical possibilities. Nordhaus's ideas[25] are closely connected with the analyses mentioned above; according to his model, the nature and rate of technical development are not factors of a non-economic nature, but are dependent on the financial resources available for research and education. The function $Q[A(t), B(t)]$ is, then, not given but can change depending on the allocation of manpower to technical development. Shell has developed a model, in which the act of invention is treated as a production process which can be described by an ordinary production function.[26]

Apart from the models of Nordhaus and Shell, which show that a concrete meaning can be given to the endogeneous character of technical change, there have been some other impressive results in this field. Thus, Arrow[27] regards technical change as a learning process and, concentrating on labour-augmenting technical change embodied in capital goods, proposes the production function:

$$Q(v,t) = F[K(v,t), A(v).L(v,t)]$$

The term $A(v)$, which represents technical change, now depends on the accumulated total investment:

$$G(t) = \int_{-\infty}^{t} K(v,t).dv$$

which determines whether more experience is acquired, leading to a steady increase in the productivity of labour. Arrow himself says that, in his model, the learning process[28] is a by-product of production. Education and research directly bring about an expansion of technical possibilities. Fu-Sen Chen[29] further developed Arrow's learning model in an interesting way on the basis of the curve for the technical possibilities proposed by Kennedy and others. Arrow's learning model can also be represented by intertemporal production functions.[30]

Following in Schultz's footsteps,[31] Uzawa[32] has constructed a model in which the rate of technical change is not given exogeneously, but is a function of the level of employment in the education sector of the economy. On the basis of the production function $Q(t) = F[K(t), A(t).L_p(t)]$, where $L_p(t)$ denotes employment in the non-education sector, the percentage change in the term $A(t)$ is ascribed to the percentage change in employment in education $L_E(t)$. This general approach has been refined by Phelps, who takes the diffusion of technical knowledge into account as well, and by W. Lachmann.[33] In Phelps' model, the relative change in the term $A(t)$ depends on the level of employment in research.[34] In an interesting model, Eisen has introduced a competitive research sector, financed by taxes on output in the production sector. In view of this budget constraint, the research sector maximizes its output.[35]

A common feature of all these macroeconomic models is that the rate of technical change is no longer an exogenous quantity in the production function, but is explained by the way in which scarce resources with alternative applications are utilized in the economic process, i.e. by endogenous factors, or, in Eltis' words, 'the rate of technical progress will be influenced by some aspects of an economy's investment activity'.[36] Lucas goes so far as to regard the entire technical change as the withdrawal of means of production from production for purposes of what he calls 'technological investment'.[37] A subject currently under investigation is how the hypotheses about the allocation of scarce resources affects the growth paths of production and the factors of production, the central issue being the derivation of the conditions for equilibrium growth.[38] A critical point here is that the significance of neutral technical change in Harrod's sense decreases if technical change is viewed as an endogenous phenomenon.[39]

An important specific problem arising in the case of technical change embodied in capital goods concerns the optimal replacement of capital goods,[40] and we must mention here Adachi's refined analysis,[41] where a two-sector putty-clay model is constructed in which the physical durability of capital can be different in the two sectors, depending on the expected prices of capital goods and the production elasticity of labour.

Although the results are still somewhat rough and ready on account of the macroeconomic nature of these models, one can already draw the important conclusion for economic policy that the rate of technical development is connected with investment in education and research.[42] Empirical investigations are necessary to determine how close this connection is, and a more detailed microeconomic analysis is needed as the basis for more specific statements about the effect of target-oriented education and research on the nature of technical development.

10.4 TECHNICAL CHANGE AND THE LABOUR MARKET

In the growth models that incorporate technical development a full-employment labour market is usually assumed; the main aim of the analysis is then to find out whether an equilibrium growth[43] of production and the factors of production is possible. The answer is generally 'yes' if a neutral technical change in Harrod's sense is assumed. A constant ratio between output and capital, either immediately or in the long run, is possible only if the technical change is exclusively labour-augmenting. If it is, then the existence of an equilibrium growth rate in the system can be demonstrated both in the case when the technical possibilities are described by a production function in which factor substitutability is assumed, and in the case where there is only one efficient combination of capital and labour at any given moment.

In the first case, the price mechanism ensures that the variable capital coefficient closely follows the exogeneously determined growth path of the working population. The full utilization of the factors of production is then just as possible with a neutral technical change in Harrod's sense as a given, constant state of technology. Furthermore, Solow and others[44] have shown that an equilibrium growth rate also exists in the second case, i.e. when there are constant technical coefficients. When labour and capital are not substitutable, the equilibrium growth rate of the system is determined by the rate of growth n of the population and the rate λ of neutral technical change in Harrod's sense. The growth rate of the stock of capital goods must then be equal to $n + \lambda$. In the case of constant technical coefficients and a Harrod-neutral technical change, full employment requires a certain increase in the stock of capital goods, just as the full utilization of the labour force at a given state of technology requires a stock of capital goods of a certain size.

The next point is whether the existence of an equilibrium growth rate means that the postulated economic system will necessarily grow in an equilibrated manner. It is clear from the outset that, in the case of Leontief's production functions, equilibrium growth is a fortuitous phenomenon. There is no endogenous mechanism that guarantees that the growth rate of capital is equal to the growth rate of the working population, measured in efficiency units. If, in this situation, which resembles Marx's approach, not enough capital is accumulated, then government intervention is needed to prevent structural and technological unemployment.

Nor is short-term unemployment ruled out in a neoclassical situation, since the moves towards the path of equilibrium growth take time. Thus, R. Sato[45] has pointed out that the period of adjustment or adaptation can be so long, or the rate of substitution so slow, that labour as a factor of production is under-utilized for quite considerable periods. Sato goes so far as to argue that, in many cases of a certain type, the adjustment is so slow that constant technical coefficients can be assumed for practical purposes. On the other hand, another author with the same name, K. Sato,[46] argues, with the aid of Phelps' model,[47] that the adjustment is considerably faster if technical change is embodied in new capital goods. The older capital goods are then written off and are replaced sooner, and the size of the stock of capital goods adapts itself more readily to the value of the capital coefficient along the equilibrium growth path. However, this result depends to a large extent on the Cobb-Douglas production function, as a result of which Harrod-neutral technical change, which is necessary for an equilibrium growth, is accompanied by neutral technical change in Solow's sense embodied in capital goods, so that more efficient machines are introduced.[48] At a given propensity to save, the capital-augmenting character of technical change ensures that the under-utilization of labour disappears more quickly than it does when only neutral technical change in Harrod's sense is involved. As mentioned before, only neutral technical change in Harrod's sense is compatible with equilibrium growth models based on other than the Cobb-

Douglas production function. Technical change is then exclusively labour-augmenting and therefore not capital-augmenting, so that, given a certain rate of saving, full employment is reached less quickly than in the previous case.

The time has now come to examine in more detail the compensation theory, which was discussed in broad outlines in Section 2.8. It was pointed out there that the significance of compensation should be judged by the endogeneous compensating factors, i.e. factors that are connected with the introduction of a new technology. It can now be stated in general that the extent of compensation in the strict sense depends on (a) the extent of substitution that is technically possible according to the production function, (b) the extent to which the possibilities of substitution are dependent on market forces and (c) the nature of the technical change.

In the case of Harrod-neutral technical change, one cannot expect much from compensation in the strict sense. Even if wages drop, due to perfect competition on the labour market, this does not, in the case of fixed production proportions, lead to compensating opportunities for work, since no alternative technique of production is available. It is admittedly possible that the higher profits are invested, so that a new demand for labour arises via the accumulation of capital, and endogeneous compensation is produced. If we assume that labour-augmenting technical change hardly lowers wages because of the workers' control of the labour market, then compensation depends entirely on the accumulation of capital, which takes place in any case independently of technical change. This situation, in which the compensation relies on exogeneous factors, closely resembles Marx's case—assuming, as in his analysis, wages cannot drop below the subsistence level.

Let us next assume that there is factor substitutability and that, owing to the drop in wages, an alternative technique is chosen. Neutral technical change in Harrod's sense then leads in the long run to the re-employment of the displaced labour force by an endogeneous mechanism. However, it is questionable whether a flexible price mechanism can be relied on, especially as wages are in practice usually downwardly rigid. If they are not flexible, full employment depends, particularly in the short run, on the exogenous accumulation of capital. The adjustment is somewhat faster only if the Cobb-Douglas production function describes the real state of affairs, since then neutral technical development in Harrod's sense means that a capital-augmenting process is also in operation. As we have seen before, neutral technical change in Harrod's sense is an essential condition for the derivation of the equilibrium growth paths of production and the factors of production. If this type of technical change in the narrow sense predominates in practice, there is not much compensation in the strict sense, particularly if the market is monopolistically competitive at the same time. If exogenous capital formation is insufficient, stagnation can be expected for quite considerable periods.

Since this generally does not happen in reality, it is logical to turn our attention now to neutral technical change in Solow's sense, less popular in theoretical studies. Although this makes it impossible to derive the equilibrium growth paths from the macroeconomic quantities, except in the special case of the Cobb-Douglas production function, the capital-augmenting character of this type of technical change makes it eminently possible for compensation in the strict sense to assert itself. This is because technical change then implies a direct saving of capital, so that a clearly endogenous compensating factor is in operation. If workers are displaced by capital-augmenting techniques, capital is liberated, so that the workers can be re-employed. With fixed proportions in production, at least partial compensation will arise (where the wages can also remain the same), depending on the rate of capital-augmenting technical change and on the technical coefficient of labour. Full compensation can be expected even in the relatively short term when the factors of production are substitutable, since then the economy can always operate with a new combination of these factors; the result is full utilization of the liberated capital and the available labour, without a drop in wages being necessary.

Our analysis has so far been confined to some main types of neutral-technical change. The detailed analysis of the effect of non-neutral technical change on the labour market is hampered by the great variety of situations that can be distinguished in part due to the various formulations of the production function. However, the same argument is generally valid—that capital-saving technical change is a powerful endogenous compensating factor, the efficiency of which increases with the possibility of substitution. Technical change, which in fact 'should be viewed as a basket of neutral and non-neutral components',[49] contains so many capital-saving elements that at least partial compensation in the strict sense can be expected, even when wages are downwardly rigid.

It has so far been left indeterminate whether or not technical change is embodied in new capital goods. With endogenous compensation, this question is of secondary importance. If, however, technical change is embodied in new capital goods, the application of new methods presupposes that the new machines have already been made, which is possible, as we have seen (Section 2.8), only if there is additional capital formation, so that compensation is then based on exogenous factors. It should be noticed that if technical change is embodied in new capital goods there are some special demands placed on those workers not made redundant. In the absence of adjustment and retraining, redundancy may arise, accompanied by a demand for more skilled manpower. This means that people can safeguard themselves against unemployment by undergoing retraining of the type required by the technical change. Automation, which partly consists of the application of labour-saving methods embodied in new capital goods, makes existing occupations obsolete but also creates new ones at the same time.

Technical change can affect not only the factor prices, but also the prices of goods and services. Though this is amenable to analysis if a neo-classical situation is assumed with a large number of entrepreneurs striving to maximize their profit,[50] the discussion becomes more difficult the closer we get to reality. Does modern big business, operating as it does in an oligopolistic market, pass on the benefits of technical development to the customers in the form of price reductions? If it does, the increased demand for the products which results can be seen as an endogenous compensating factor, should the new technique reduce the number of jobs. If the prices are indeed downwardly rigid, then the higher profits lead to investment, which can in turn create new jobs. The complicated connections between technical change and oligopoly will be examined in the next chapter, so we will limit ourselves here to saying that compensation in the strict sense probably results more from the investment of the extra profits due to the new techniques than from an increase in demand caused by a drop in prices. Compensation in the broad sense can admittedly arise from an increase in the demand, independently of technical change, but this factor is just as exogenous as that accumulation of capital which is not connected with the application of new techniques.

If technical development itself is regarded as an endogenous process which is considerably influenced by research and development, then labour-saving methods can naturally also result from R and D. The possibilities of compensation can then be discussed in exactly the same way as before, except that here one has to consider to what extent the engagement of some people precedes the dismissal of others. To ask whether compensation in the strict or broad sense is involved vis-à-vis, for example, the jobs created in the field of education in the past few years is rather like asking whether the chicken or the egg came first.

We have therefore seen that neutral technical change in Harrod's sense is perfectly compatible with equilibrium growth but leads to a pessimistic conclusion about compensation in the strict sense. But we have also seen that there are several other types of technical change—neutral and non-neutral, exogenous and endogenous, embodied and disembodied—which can lead to an optimistic conclusion. However, if we stress these types with a view to exploring the theory of compensation, we lose to some extent the framework of the modern theory of growth.[51] In this way, the equilibrium growth in theory gives way to the non-equilibrium growth of employment in reality. There are different growth rates in the various sectors of the economy, nor are such growth rates constant for ever. Akerlof and Nordhaus,[52] not without reason, have said that economists who deal with long-term development 'might begin to think in terms of models which show unbalanced growth'. Technical development, by its very nature, alters the situation just as one is about to pronounce it an equilibrium. The usual picture is of redundancies in one sector and unfilled vacancies in another. While this can lead to difficult problems of adjustment temporarily, the question is whether compensating employment is created overall. The answer is that it is, by virtue of the heterogeneity of technical development which underlies the somewhat spasmodic course of development; however, retraining is necessary, for it speeds up the transition from old techniques to new ones.[53] And as the endogenous character of technical development itself grows in importance, one can begin to control and regulate, depending on the strength of the endogenous factor. The labour market can retain an equilibrium structure, thanks to the non-equilibrium character of technical development.

Our optimistic view contrasts with Kalecki's pessimistic one, based on his analysis of the dynamics of capitalist economy. This author first developed a theory of the pure business cycle in a stationary economy and then introduced trends into it, separating the short and long-term effects. He argues that innovations tend to increase the level of investment in the long run, and this determines the long-term trend. The stream of inventions makes new investment attractive and may therefore be compared to a steady increase in profits; inventions transform a static system into one with an upward trend. Stagnation in the later stages of capitalist economies is due at least partly to the decline in the intensity of innovative work, which can be explained firstly by the diminishing importance of the discovery of new raw materials and secondly by the increasingly monopolistic features of capitalism militating against the applications of new inventions.[54] These factors may cause a rise in unemployment in the long term. Later Kalecki acknowledged that by separating short-term and long-term effects he missed 'certain repercussions of technical progress which affect the dynamic process as a whole.'[55]

The view that trends and the business cycle are closely intertwined has also been put forward by Akerman.[56] Kalecki has developed a refined model in which investment is related to trend and business cycle components, both of which are influenced by technical change. It follows from Kalecki's analysis that the rate of growth of the whole system is determined not only by the coefficients of the equations for the period under consideration, but also by past economic, social and technical developments. Besides, the use of equipment and labour is influenced by the degree of monopoly and the maximum growth rate of a time function that reflects past economic, social and technical developments. Certain combinations of the values of coefficients describing these factors imply stagnation, which in Kalecki's view is due not to a lack of inventions or investment opportunities, but to an inherent failure of the capitalist system to exploit all the possibilities—although quite apart from this, the system will not be able to generate full employment for political reasons.[57]

Although the present economic situation of most Western countries seems to support Kalecki's predictions, we maintain our optimistic outlook. Owing to the heterogeneous character of technical development, the increase in incomes and particularly in real wages will decide which of the technical possibilities will be used. Hicks says that we shall not emerge from a slump until we accept a drop in real wages in most jobs.[58] The unemployment situation is likely to improve as soon as the income distribution is considered once again to be at least partly a reflection of productivity rather than the other way around.[59] This implies that the present plight of economies is to a large extent the result of bad policy and not of internal and technical characteristics of the capitalist system. This judgement does not contradict the impression that innovative activity is much more intense during a boom than during a slump. It has even been suggested that the Kondratieff cycle can be fully explained by technical change. Although such an explanation, based on a single cause, seems highly unlikely in view of historical factors acting over long periods, it is beyond doubt that both inventive and innovative activity are subject to the ups and downs of general economic life in the short and the long run.

10.5 EMPIRICAL RESULTS

Empirical testing is a necessary complement to the construction of theories, and it sobers the excesses of pure speculation. Empirical testing can never *prove* the general validity of an economic theory, since further observations may contradict the conclusions of the theory or the empirical data may turn out to be explained better by another theory. In principle, empirical tests tell us to what extent the theory conflicts with reality; they generally lead not to a final assessment of existing economic theories, but to the recognition of new problems and the formulation of new theories. Econometrics deals with the empirical testing of economic models and has often been instrumental in the improvement of economic theories, which always have a provisional character. It can be argued, with Popper,[60] that theoretical economics becomes an empirical science only if its statements can be tested and refuted by experience, in which case the theory is said to be 'falsifiable'. The test of falsifiability precedes empirical testing, because it concerns the logic of the theory. According to Popper's criterion, the empirical testing of statements derived from a theory also involves testing the assumptions of that theory; but the question of whether there is also a place for the direct and separate testing of hypotheses introduced into the theory cannot be discussed here.[61]

Falsifiable economic theories generally need to be further specified before they can be subjected in practice to empirical testing. Apart from stochastic elements, additional hypotheses must generally be introduced.[62] There may also be statistical difficulties in the measurement of the operational quantities and concepts, as for example in the case of elasticity of substitution. Finally, one is faced with the standard econometric problem of the identification of the relations involved in the theory.

It is under very special conditions, if at all, that the individual relations of a model can be tested, i.e. that the parameters of the relations can be identified. Under the influence of a whole complex of relationships, some equilibrium values exist in each period which are in principle reflected in the statistical material so that the connections between these values can be established. The underlying set of relations is not then amenable to direct testing and identification. One can easily get the impression that an econometrically estimated parameter corresponds to a coefficient that features in the model, as it were, in the same position. However, the propensity to consume, for example, that can be determined statistically from national consumption figures and from the national income is not the same as the propensity to consume used in Keynes' *ex ante* consumption function. It is not least because of the problems of identification that econometric studies raise new questions rather than settle old ones.

Empirical research concerning technical change can be roughly divided into macroeconomic and microeconomic studies. The former concentrate on the effects of applying new techniques, while the latter examine the causes of the expansion of technical possibilities.[63] This section deals mainly with empirical macroeconomic studies, while the more microeconomic investigations are discussed in Chapter 11.

Engineering production functions, which have a microeconomic character, are an exception, however, to the above characterization. The empirical nature of these functions justifies their brief discussion at this point. Engineering production functions are constructed on the basis of information given by engineers and technologists about the purely technical aspects of production, such as Chenery's description[64] of the construction of pipelines for natural gas. It is important to realize here that these engineering production functions deal with only one aspect of production and that they reflect only the technology actually in use, so that other technical possibilities are not taken into account. On the other hand, they are eminently suitable for expressing developments in the use of techniques since there is a new engineering production function for each new technical application.[65] Finally, the empirical formulation of these engineering production functions gives an indication of the extent to which *ex post* substitution between labour and capital is possible,[66] and of the significance of changes in scale.

By focussing attention on empirical research *vis-à-vis* the consequences of technical change as reflected, for example, in the volume of production, we really lose sight of technical development in the sense of emergence of new technical possibilities. This is because the changes in the volume of production depend not only on the diffusion and application of technical knowledge, but also on changes in scale, the substitution between labour and capital due to changes in prices, the improvement of the quality of work, and relative changes between the various sectors of production. The aim of constructing econometric models here is mainly to determine the relative importance of the various factors in total output.[67]

If we start with a simple production function which involves only the factors of production, then econometric testing requires a more precise definition of this function. Once statistical studies have been made on the basis of the Cobb-Douglas production function, with an elasticity of substitution of 1, we can change over to more general production functions operating with a constant elasticity of substitution. However, even if the postulated production function fits the empirical data in all the cases, theoretical considerations urge us on to more complicated production functions that take account of technical change. If we start with neutral technical change and ascribe the unexplained part of the growth of production to non-neutral technical change, scientific curiosity urges us to examine the latter more closely. It has so far been assumed that technical change increases the productivity of old and new capital goods to the same extent; we now add the refinement that technical change is embodied in new capital goods. Empirical studies concerning the theory of production and technical change thus illustrate how fruitful are the interactions between econometric research and falsifiable economic theories.

Specific statistical problems arise when technical change is measured by a productivity index. Such indices are either partial or total, the former being exemplified by the labour productivity index—a production index divided by an index of the working population. This partial index is not suitable for measuring technical change because, for example, the distinction between mechanization and technical change is obscured. A total index, involving either an arithmetic or a geometric average, is therefore used nowadays. In the case of an arithmetic average, the output is divided by a weighted sum of the indices of the factors of production, while in the case of a geometric average the data are processed logarithmically. When these total indices are used, every rise in production is explained by technical change, which thus features as a residual factor—a point that will later be examined more closely.

Owing to the heterogeneity of the production factors labour and capital, serious measurement difficulties arise in attempts to quantify these aggregates. The capital goods are installed in different years and also differ in productivity and price,[68] and thus a rather arbitrary basis of evaluation has to be introduced for the aggregation of these heterogeneous capital goods. Correction must be made for the degree of utilization of the capital goods, particularly in the case of studies involving a time series. A quality index must also be constructed for the capital goods, according to the way the technical development is viewed. There are also some problems in the measurement of labour, especially when its productivity increases in the course of time as a result of training and education.

A different problem—one of identification—arises when the econometrician abandons the simple approach of testing a single relation by correlation or regression analysis and begins to test a set of simultaneous equations, which promises a better explanation of the phenomena. The parameters of the *ex ante* production function cannot as a rule be identified on the basis of *ex post* observations. A 'production function' of a certain type, postulated .on the basis of *ex post* observations of the production and the factors of production, should be carefully distinguished from the *ex ante* production function, introduced into the theory for characterizing the technical possibilities, since the *ex post* relationship depends also on some non-economic factors. It is a complicated econometric problem to decide whether the various equilibrium positions found in the analysis of a time series or in a cross-sectional study can be explained by a shift in the production function or by a movement along the production function. These complications arise particularly when (i) non-neutral technical change is also taken into account, (ii) technical change can be embodied in capital goods, and (iii) scale effects are in operation. It is then highly questionable whether the estimated parameters, such as the elasticity of production and substitution, the scale coefficients and the criteria for embodied and disembodied technical change, correspond to their counterparts selected *ex ante* which feature in the economic model proposed.

As mentioned before, Cobb and Douglas were the first to make an ambitious attempt to construct a production function in an empirical manner. In this framework, the Cobb-Douglas production function has an *ex post* character, so that it sheds no light at all on the underlying system of relations which are supposed to explain the set of equilibrium values of production and the factors of production. Menderhausen pointed out a long time ago that there was no possibility of testing a production function, and that it was the empirical coefficients that expressed technical change.[69] Douglas, who contemplated switching to another field of enquiry[70] because of the fierce criticism he received, in fact continued along the same path, as can be seen from his article published in 1948.[71] Prompted by a suggestion from Durand,[72] he no longer started with constant returns to scale, so that increasing returns to scale no longer affect the general yardstick of productivity; however, in other respects he follows the process of establishing a direct connection between production and the factors of production by means of the correlation method. The field of investigation is extended to cover not only the analysis with the aid of a time series, but also cross-sectional studies of the various sectors in the same period. Both approaches show that the hypothesis of constant returns to scale fits well with the empirical data, and the estimated production elasticity of labour closely corresponds to the share of the wages in national income.[73] However these results are not convincing since the author ignored technical change.

In a study[74] published in 1942 Tinbergen added a trend term to the Cobb-Douglas production function which reflects neutral technical change. Starting with values of $\frac{3}{4}$ and $\frac{1}{4}$ for the production of elasticities, Tinbergen estimated this trend term on the basis of data for the UK, the USA, Germany and France, covering the period 1870–1914. Tinbergen's work represents a big step forward from Cobb and Douglas's empirical studies, but his *a priori* approach to the values and the sum of the production elasticities is open to serious criticism.

After Marschak and Andrews[75] pointed out the need to base empirical verification of production functions on a set of simultaneous equations, Solow did some pioneering work on this requirement (some economists, however, still resort to the traditional, direct testing of the production function). Solow[76] has assumed constant returns to scale, neutral technical change in Hicks' sense, the existence of macroeconomic production function and the validity of the theory of marginal productivity. His data, which refer to the non-agricultural private sector in the USA in the period 1909–49, are best analyzed with the Cobb-Douglas production function. Solow's arithmetic index of technical change increases by 1 per cent a year in the first half of this period and by 2 per cent in the second half. This means that the doubling of production in the forty-year period can be attributed almost 90 per cent to technical change and only about 10 per cent to capital formation. Massell[77] has obtained practically identical results for the industrial sector in the USA for the period 1919–55. A comparable study has been done by Archibald[78] for Greek manufacturing industry between 1951 and 1961, and by McLean for the agricultural sector in Victoria,[79] Australia.

These investigations by Solow, Massell and Archibald have not been undisputed. Criticism has been levelled at the statistical data and the method employed, and attention drawn to the hypothetical nature of neutral technical change not embodied in capital goods, the assumption of constant returns to scale, and the restriction of the investigation to the case where the elasticity of substitution is 1; besides, the rise in productivity attributed to technical change may be due to changes in the pattern of production, better education and training, and a better social and organizational climate.[80]

Using less refined models, 'Abramowitz's residual', which comprises, besides the inputs of the factors of production, all other factors that raise output, is obtained by eliminating from a time series of production the contribution of labour and capital. This has been done by Abramowitz,[81] Kendrick[82] and Denison,[83] following in the footsteps of Kuznets.[84] If we try to assess the importance of technical change in this way, we are often forced to introduce rather arbitrary assumptions about the quantitative contribution of factors other than labour and capital. These can in principle be avoided by constructing a model of economic change which takes account of the various factors that influence the level of production and which contains sufficiently refined hypotheses about technical development. Tackling the empirical data with a rough model can easily lead to a serious overestimate of the contribution of technical change, since in such a model there is no room for the effects of other factors. Another danger of this approach is that the basically very complicated process involving interacting factors is oversimplified. It is therefore encouraging to see that since the early sixties more sophisticated ideas about technical change have been balanced by more refined econometric models.[85]

Brown and Popkin,[86] who distinguish between neutral and non-neutral technical change, have made a significant contribution in this field. Though they use a Cobb-Douglas production function, they do not assume constant returns to scale. Using a statistical method, they distinguish between three periods with only neutral technical change in the non-agricultural private sector of the USA between 1890 and 1958. In these periods the production figures can be explained by the labour and capital inputs, neutral technical change and returns to scale. As the following figures show, the parameters of the Cobb-Douglas production function change from one period to the next, which is due precisely to non-neutral technical change.

	Production elasticity of labour α	Production elasticity of capital β	$\alpha + \beta$
1890–1918	0·97	0·49	1·46
1919–37	0·43	0·60	1·03
1938–58	0·51	0·53	1·04

A major objection to Brown and Popkin's method is that it uses a production function that implies a constant elasticity of substitution of 1. We must therefore examine production functions without this restriction.

In the study by Arrow, Chenery, Minhas and Solow, mentioned earlier, the elasticity of substitution is estimated at 0·57 on the basis of data which Solow also used for his 1957 study. The annual increase in productivity is estimated at 1·8 per cent and the technical change is neutral. In the cross-sectional investigations, furthermore, the elasticity of substitution varies between 0·72 and 1·01, according to the sector examined.

According to Kendrick and Sato's[87] index, neutral technical change in Hick's sense was 2.1 per cent per annum in the USA between 1919 and 1960, while the elasticity of substitution in the CES production function was 0·58. Brown and de Cani[88] have estimated the values of neutral technical change, the returns to scale, and the elasticities of substitution for the three periods 1890–1918, 1919–37 and 1938–58. Ferguson has carried out several studies on the CES production function, and in the first of these he says that strictly speaking, the empirically observed combinations of the quantities of labour and capital leading to one and the same value added must not be confused with the production function that describes the technical possibilities of mutual substitution between labour and capital.[89] Ferguson later returned to this difficult problem of identification in his other studies and concluded that over 90 per cent of the growth of production in 1929–63 can be explained by neutral technical change in Hicks' sense and the same applies to the period 1948–63 if neutral technical change in Harrod's sense is assumed. The elasticity of substitution for the CES function is less than 1.[90] These results essentially confirm Solow's findings published in 1957. Investigating technical change in a large number of sectors for the period 1949–61, Ferguson concluded that the technical change was neutral and capital-using.[91] The American David and the Dutchman van de Klundert have made an ambitious attempt to explain the growth of production in the American private sector between 1899 and 1960 with the aid of a CES production function operating with non-neutral technical change, and they obtained the figure of 1.85 per cent for the annual technical change, which on average is more labour-saving than neutral in Hicks' sense.[92] Both these results and Ferguson's have been criticized from the economic point of view, in particular the identification of the production function and the arbitrary types of technical change based on time series.[93] The hypothesis of a constant elasticity of substitution has been criticized by Revankar, who has carried out some investigations[94] on the basis of a VES function, which are otherwise very similar to those of David and van de Klundert.

The empirical research discussed so far has been based on the hypothesis that technical change is not embodied in new capital goods. Even so, the interpretation of the results involves serious problems, and the econometrician dealing with the effect of technical change on economic growth is faced with the difficult task of distinguishing and defining the *ex post* and the *ex ante* production functions, effecting the necessary corrections for the degree of utilization of the factors of production, assessing the influence of the increase in scale on productivity, and separating technical change into neutral and non-neutral components. There are even more complications if it is assumed that the technical change is embodied in new capital goods. The identification of the parameters of the production function that give an account of technical change is then made more difficult by two things in particular. If technical change is embodied exclusively in new capital goods, splitting the growth in the stock of capital goods into a quantitative and a qualitative component presupposes a knowledge of the rate of embodied technical change, which is precisely the subject of these investigations.

When, in the explanation of the growth of productivity, embodied and disembodied change cannot be assessed in advance, we are faced with an almost insoluble problem of assignment in connection with 'Abramowitz's residual'.[95] The assumptions that the level of investment as such raises the level of technical knowledge, and that technical change is embodied in labour as well as in capital make the proposed picture more realistic, but at the same time they make the validity of the estimation of the parameters and the observed relations more dubious.

It is hardly surprising, therefore, that some economists prefer to discard the hypotheses of embodied technical change. Thus, Denison[96] believes that the significance of technical change embodied in capital goods depends exclusively on the age-distribution of the total stock of capital goods, which in his opinion is only subject to very small variations. In a detailed study, Jorgenson[97] comes to the conclusion that a model of embodied technical change cannot be distinguished from one of disembodied technical change on the basis of empirical data. Solow, who, as Smithies says, is 'like a Pied Piper who can play different tunes',[98] observes that the macroeconomic empirical data can be explained just as well by the one type of technical change as by the other.[99] This conclusion, which seems at first rather strange, should be examined in conjunction with the *ex post* character of the production function that describes embodied technical change (cf. Section 10.2). Empirical testing has no bearing on the *ex ante* production function in one case and on the *ex post* production function in the other, but in both cases, on the equilibrium positions, which implies a choice that is only partly determined by the technical possibilities. The solution of the econometric identification problem may mean that the *ex ante* production function operating with disembodied technical change is distinguished more sharply from the production function operating with technical change embodied in capital goods.[100]

While we may hope for refinements in econometric research, what are the results so far obtained with the aid of econometric models operating with technical change embodied in capital goods? In Solow's study,[101] in which disembodied technical change does not feature, a value of 0.025 is assigned to λ, which means that embodied technical change raises the production by 2.5 per cent a year; this high value is partly due to the fact that disembodied technical change has been ignored. While a shifting production function is assumed in this study, as in the study published in 1962,[102] Griliches[103] takes a stable production function and tries to explain the changes in the production by the changes in the quality and quantity of the factors of production and by the changes in scale. In this case, the 'Abramowitz's residual' is the sum of the errors involved in the measurement. Griliches succeeds to a great extent in explaining agricultural production in the USA in the period 1940–60 by these factors. He therefore concludes that technical change is embodied in both labour and capital—not only in the former, as assumed by Denison, nor only in the latter, as assumed by Solow.

In the 1963 de Vries lectures Solow[104] published the results of a model of embodied technical change applied to the USA and Germany, in which the connection between λ and the production elasticity of labour and capital plays an important role. In the following year, Nelson[105] published an excellent survey in which he makes a clear distinction between quantitative and qualitative changes in the stock of capital goods, and—in the case of qualitative changes—differentiates between improvements and changes in the age-distribution of the stock. Furthermore, he introduces a parameter for the improved quality of labour, due for example to education and training. He uses a Cobb-Douglas production function to express the embodiment of technical change in labour and capital. His most important conclusions are that it is wrong to suppose that the various factors which determine the total productivity are independent of one another, and that the changes in the nature and size of the stock of capital goods, in the working population and in the scale effects should be seen as interacting phenomena. If technical development is regarded as an endogenous process, we cannot estimate the growth rates by regression analysis in which these quantities are taken to be constant. Gahlen[106] reaches a similar conclusion on the basis of a very detailed study of the growth of production in Germany, in which he uses Cobb-Douglas and CES production functions to examine neutral technical change.

Intrilligator[107] has developed a model which includes both types of technical change. His conclusion for the USA in the period 1929–58 is that the best result is obtained when both embodied and disembodied technical change are taken into account, raising the production annually by 4 per cent and 1.67 per cent respectively. Szakolczai and Stahl[108] have developed this work by discarding constant returns to scale. Phelps and Phelps have similarly included both types of technical change in their econometric investigations[109] (in which they use Samuelson's wage-interest frontier).

Jorgenson and Griliches' impressive study,[110] in which the growth of production is explained almost entirely by the factors of production, is diametrically opposed to the empirical studies in which 'Abramowitz's residual', ascribed to technical change, plays an important part. Jorgenson and Griliches corrected the errors of measurement attributable to the aggregation of consumer and capital goods, the changing degree of utilization of capital, and the aggregation of capital goods and units of labour. They thus estimated that 96.7 per cent of the growth of production in the USA in the period 1945–65 is explained by growth in the factors of production. These interesting findings largely result from the assumption that labour and capital must be measured according to their contribution to production. A capital good that, owing to disembodied technical change, makes twice the contribution to production at time t as at previous time $t-1$ is counted twice. In this way technical change is swallowed up by the stock of capital goods and the working population. This approach gives a better account of exogenous technical change, but is open to the criticism that can be levelled against any quantification of a qualitative process. It is more natural for empirical research to examine the endogenous factors that can explain technical change. An example is Sheshinski's investigation[111] which tests Arrow's model of learning. Sheshinski has found that the productivity of labour is greatly affected by investments, which are a measure of the acquisition of more experience. However, he combines this affirmation of Arrow's theory with the criticism that technical change embodied in capital goods was ignored. A recent attempt to test the empirical relevance of the endogenous component of technical change has been made by E. K. Y. Chen.[112]

There have been some further attempts to unravel 'Abramowitz's residual' by resolving the empirical observations of technical change into various neutral and non-neutral types.[113] However, the problems of using statistical and econometric methods, the uncertainties afflicting the various specifications of economic models, and the qualifications which must be made about the results of empirical investigations mean that our understanding of the complicated relations between growth, technical development and the factors of production is far from perfect.

10.6 CONCLUSIONS

Solow has pointed out that a position of equilibrium growth is 'not a bad place for a theory of growth to start, but may be a dangerous place for it to end'.[114] This is particularly true vis-à-vis technical change. Equilibrium growth is the exception rather than the rule, even if technical change in the strict sense is confined to an increase in technical knowledge and the analysis restricted to the macroeconomic case. Only one of many views of technical development is compatible with the derivation of a set of equilibrium growth paths of production and the factors of production. If one operates simultaneously with neutral and non-neutral, embodied and disembodied, endogenous and exogenous technical change, the position of equilibrium growth cannot be derived. If the analysis is repeated on the microeconomic plane, the heterogeneous character of the process of growth can be ascribed to the variety of technical change. From one point of view equilibrium growth and structural change represent statics and dynamics respectively.[115]

In the case of technical development in the broad sense—an increase and change in the technical possibilities—the above conclusion is reinforced, since scale effects, investment in education, further division of labour and improvement in living conditions generally do not exhibit the uniformity necessary for equilibrium growth. Technical development in the broad sense (which closely resembles Marx's concept) makes macro-economic growth something like an eventful play, full of surprises and unexpected turns, sometimes spontaneous and sometimes induced, with the nature and volume of production uncertain.[116]

Empirical analysis of the quantitative contribution of technical change to growth are indeterminate, partly because of the diversity of technical development and the difficulty of separating the development of technical possibilities from the expansion of technical knowledge, which are not only to some extent interdependent but also hardly, if at all, directly observable. Empirical observations in this field form a complex that cannot be unravelled without introducing arbitrary assumptions.

A similar problem of assignment crops up in the connection between employment and technical change. Here the aim of empirical research is to allocate the total demand for labour between the factors which determine this demand, such as changes in scale, change in the relative prices of the factors of production, the effect of trade cycles, seasonal variations and technical development (which needs further definition). Operating with a system comprising an industrial, an agricultural and a services sector, Kuznets[117] has found that during the growth of developing countries, the proportion of the working population employed in agriculture has sharply decreased, while the proportion of the working population employed in industry has risen. The greatest increase in the number of jobs occurs in the services sector. It can be seen from H. Reinoud's investigations[118] for Holland and France, which are very similar to those of Kuznets, that technical development lowers the demand for unskilled manpower and raises the demand for skilled or qualified manpower. On the assumption that the level of education and training is steadily rising, there are no grounds for the fear that technical development will produce general unemployment. A thorough empirical survey by an American team[119] similarly indicates fundamental shifts in the demand for various types of manpower, but does not predict a permanent reduction, due to technical development, in the total demand for labour. The difficulty in interpreting these results lies in deciding how easily technical development can be seaparated from the other factors that determine employment, in particular to what extent full employment depends on the accumulation of capital occurring independently of technical development. Further empirical research is needed here to verify the compensation theory in the strict sense. Such investigations could be based on multi-sector models which take into account the interactions between technical development and capital formation.[120]

The construction of the Cobb-Douglas production function was closely linked with the explanation of income distribution, although this production function was later incorporated as an independent entity into the theories of production and growth. We ought therefore to discuss briefly the connection between technical change and income distribution.[121] The concept of neutral technical change has been introduced into the theory with a view to elucidating the role of labour and capital in the process of income distribution. Cobb and Douglas started with the simple concept that the constancy of the share of wages in the national income could be explained by an exponentially linear production function in which no scope was allowed for technical change. However, this simple assumption can no longer be maintained. The evolution of marginal productivity in the course of time depends first on the shape the production function is assumed to have and then on the type of technical change introduced. Thus, if a technical change embodied in new capital goods is assumed, for example, the marginal productivity of workers using old capital goods will not be equal to that of workers using new capital goods until a redistribution of the workers has taken place (no such rearrangement is necessary with disembodied technical change). Since the two cases involve different manpower demand functions the level of wages and thus the income distribution will also be different. If non-neutral technical change is involved and constant returns to scale are again assumed, income distribution depends entirely on the specific content of this type of technical change.[122]

Cobb and Douglas' original study was simplistic not only in disregarding technical change but also in its assumption of constant returns to scale, which made it possible to explain the division of total income into a wage and non-wage component. If there are increasing returns to scale as well as technical change, the marginal productivity of the factors of production increases with the output of production, and the income distribution is no longer determined in accordance with the theory of marginal productivity. These considerations illustrate the theoretical nature of the original framework; in a macroeconomic context increasing returns to scale cannot be simply dismissed at the outset.

We should treat carefully an analysis which, while leading to interesting analytical results, deviates considerably from reality. Explaining the income distribution in a purely macroeconomic framework with the aid of a shifting production function that treats technical change in one way or another involves the risk of constructing a system which, though logically self-consistent, is remote from reality. Ignoring market power leads to the belief that technical change lowers prices of goods, so that the real income of employees increases. But entrepreneurs' control of the goods market and workers' control of the labour market may mean that technical development produces higher monetary wages and higher prices in a ratio such that real wages increase. This connection between technical development, inflation and income distribution is not amenable to a purely macroeconomic analysis; the latter must be complemented by an examination of the way the incomes are negotiated by employers' and employees' organizations at the level of the firm and the industry.

It is not only the consideration of income distribution which requires a return to microeconomic analysis: improving our understanding of the causes of technical development and of the factors that govern the choice from among technical possibilities also requires an examination of the connection between technical development and the type of market competition. A macroeconomic analysis of technical change and scale effects must be mainly theoretical; but it becomes considerably more meaningful at a microeconomic level. The discussion of the labour market in a macroeconomic context lacks substance, whereas a complementary microeconomic analysis can illuminate the effects of a technical change on the employee, the firm and the entire industry. In short, while a macroeconomic analysis is useful for a preliminary sketch of the effects of technical change a more profound understanding of technical development must ultimately be connected with human activities centred in the production process.

11 Technical change and monopoly power

The fundamental impulse that sets and keeps the capitalist engine in motion comes from the new consumers' goods, the new methods of production or transportation, the new markets, the new forms of industrial organization that capitalist enterprise creates.

Schumpeter[1]

11.1 INTRODUCTION

The macroeconomic literature so far discussed portrays technical change as a predominantly mechanical process, independent of human will and behaviour. Even when technical change is treated as endogenous from an analytical point of view, firms are assumed to have little or no latitude in matters of this sort. If macroeconomic ideas about the nature and rate of technical change are expressly or tacitly based on perfect competition, it is easy to give the impression that the entrepreneur's choice of technique emerges from simple behavioural postulates. It is assumed that the individual supplier has very limited market power and that entrepreneurs behave in a passive manner, so that particular and specific market results can be predicted.

This assumption of passive behaviour is at variance with the actively competitive behaviour which has traditionally been regarded as a driving force in economic life. Active competition between producers increases the efficiency of production and encourages the renewal or updating of equipment and the development of new products. Competition creates a social and psychological climate in which growth and change flourish in society. In this picture of competition, those involved in market transactions are not passive, but instead pursue a dynamic company policy. In this way, they have enough latitude to choose from various courses of action when formulating their policies, so that their decisions display a complex pattern and cannot be readily predicted. With technical developments and the choice of their applications, too, different results can be envisaged, reflecting the variety of circumstances and human behaviour. The latitude firms have in this respect results from their market power, and the nature and extent of this power and the effects of exercising it are so varied in reality that one should abandon any attempt to find a direct, straightforward connection between a certain market form or structure and a certain market performance.

To unravel the complex relationships between market structure, market behaviour and the applications of technical developments, we now return to Schumpeter, who, in his description of capitalism, discussed in detail the dynamic nature of competition which constantly renews the economic structure from within. This process of creative destruction must be judged by its long-term results. With a certain combination of production methods, products and market structure, price competition is less important than actual or potential competition from innovation, which changes or may change production methods, products and market structures. This is not so much 'horizontal competition' among firms at the same level in an industry, as 'vertical competition' among firms operating at different levels in the industry. Less significance is attached to the restriction of production and sales than is customary in the static allocation theory of monopoly, since the monopolist's urge to innovate is greatly intensified by the innovative competition with others outside his own industry. Since the introduction of new production methods involves risky investment, a firm must either take out patents or keep its new technical knowledge completely or partially secret, thus creating or consolidating its monopoly power. Profits arising from this relative monopoly position thus result ultimately from the acquisition, monopolization and application of new technical knowledge.

Schumpeter uses the term 'monopoly' in a relative sense, meaning the accumulation of a degree of market power such that the firm's sales are independent of the market behaviour of other companies. These monopolists have better production methods and better personnel than a 'crowd of competitors', and it is beyond doubt that 'under the conditions of our epoch such superiority is as a matter of fact the outstanding feature of the typical large-scale unit of control, though mere size is neither necessary nor sufficient for it'. This does not mean that monopolists grow rich in their sleep, because 'a monopoly position is in general no cushion to sleep on'. The introduction of new methods and new products is generally so costly and risky that 'perfect—and perfectly prompt—competition from the start', due to the unrestricted entry of new competitors to the market, can hardly be reconciled with it. The rigidity of prices offers a certain protection for firms that introduce innovations. In comparison with perfect competition, monopolistic enterprises may thus make additional profit for quite considerable periods, but this profit is the other side of the coin of that application of new techniques which does not take place under perfect competition. In this respect there is no reason to regard perfect competition as a 'model of ideal efficiency'; instead—Schumpeter argues—one should accept that large-scale production has become the most important cause of growth not in spite of, but because of, a market behaviour that 'looks so restrictive when viewed in the individual case and from the individual point of time'.[2]

Schumpeter's ideas triggered off a lively debate and this has in turn led to new theories and stimulated empirical research.[3] The 'market form' is generally taken to comprise the objective economic characteristics of the market, which in particular determine the power relationships between companies. Some of these characteristics, such as the number of firms, the nature and degree of product differentiation, and the nature of the barriers to the entry of new competitors, have already been mentioned when discussing Schumpeter's theory. The somewhat broader term 'market structure' includes vertical integration and diversification as well.

The market power of an enterprise is often measured in terms of its turnover or the number of its employees. However, these do not necessarily reflect market power accurately, for a firm that is large by these standards may possess less monopoly power in a market than a firm with, for example, a much smaller turnover. The market share, indicating size relative to the particular market, is a better yardstick than absolute size. A problem here is the definition of the market, and the theoretical studies have not produced any generally accepted criteria, because of the substitutability of goods. In each particular case, the demarcation of the market remains somewhat arbitrary.

The number and relative size of firms determine the horizontal or market concentration, namely the extent to which an industry or a market is controlled by a few companies.[4] This is often measured in terms of the market share of firms, listed according to their absolute size. The market share of the four largest firms is smaller in the case of a broad interpretation of the market than in the case of its narrow interpretation.[5] Two markets can be compared by plotting the market share along the ordinate and the cumulative number of firms (arranged according to size) along the abscissa. The resulting concentration curves produce an unambiguous statement of the degree of concentration, provided that these curves do not intersect; if they do, then the number of firms taken into account determines whether the concentration in one market is taken to be greater than that in the other one.

Firms are increasingly composed of a number of plants,[6] due to internal growth, mergers or takeovers. The degree of concentration varies, depending on whether the firms or the plants are ranked according to size, and the concentration curves for the firms differ from those for the plants to various extents.[7]

If the plants of one firm have different markets they belong to different levels of the vertical organization in the industry. The extent to which an industry is formed by firms which dominate production in this entire vertical structure is an indication of what Blair has called the degree of vertical concentration. Where a firm has plants with different vertical production structures, a degree of concentration—called conglomerate concentration by Blair—can be defined on the basis of the extent to which production in one such vertical structure is controlled by firms whose activities lie mainly in another vertical production structure.

The classifications of market forms normally encountered are based on the assumption that a market situation can be explained by all the relationships between the enterprises active in the national economy—an assumption connected with the concept of horizontal concentration. Because of the increasing importance of vertical and conglomerate concentrations, it is less and less justifiable to classify market phenomena exclusively on the basis of traditional market forms. Owing to the variety of products made by a few firms, diversification[8] and the existence of conglomerates,[9] the firms increasingly assume positions of power in different markets, thus influencing the market forms in a way that is not immediately clear.[10] One and the same firm may be, for example, a monopolist, an oligopolist or a monopolistic competitor, depending on the product considered. The intertwining pattern of the various types of competition acquires a new dimension if the firm is multinational.[11] Thus the overall monopoly power of the firm is more than the sum of its partial monopoly powers in the various national markets.

By concentrating on the relationship between technical change and oligopoly, we shall be able to illuminate the partial monopoly power of the firm in a certain market, without losing sight of the elements of the market structure that influence the overall monopoly power of the enterprise. The relation of the size of the firm to monopoly power is particularly important and is discussed in a separate section. This separate treatment is due to the fact that the other markets with which the firm has dealings are generally oligopolistic, to the fact that the various forms of concentration can be seen particularly clearly in oligopolistic markets,[12] and to the fact that there is a connection between the optimum size of the firm and the emergence of oligopoly.

As well as market structure, market behaviour is also important. Market behaviour is the firms's commercial policy which relates not only to price policy, but also to other elements of marketing. The innovative activity of the firm is usually included in its market performance, although it seems clear that the introduction of new products is an aspect of market behaviour. The *development* of new products and production methods and the introduction of new *techniques* can properly be regarded as part of market performance.

Other factors include the allocation and utilization of the means of production, the growth of production, and the effect of market structure and behaviour on income distribution.

We are quite clearly faced with complicated economic

problems here. The variety of the market structures, behaviours and performances alone complicates the theory of oligopoly and the complexities are increased by the fact that not only do the nature and extent of the monopoly power of oligopolists influence technical change, but the latter also affects this market power. Furthermore, the expansion of technical knowledge and technical possibilities must be distinguished from the applications of techniques and technologies. These complications will be tackled by prefacing our treatment of the main subject with a discussion of the generation of technical knowledge and the factors influencing the time-lag with which firms apply new technical knowledge.

Examination of the interactions between technical change and monopoly power in an oligopolistic market permits no welfare conclusions to be drawn without a further analysis using welfare economics. Nor is it possible to specify the criteria for judging the market phenomena, which would entail developing the concept of 'workable competition';[13] non-economic value judgments must be introduced to arrive at yardsticks for measuring the effectiveness of competition, which is connected with market structure, market behaviour and market performance.

11.2 THE GENERATION OF TECHNICAL KNOWLEDGE

The generation of technical knowledge consists in acquiring new ideas about production methods and goods. The acquisition and diffusion of general knowledge are not expressly directed at the production of more or different goods, and therefore only the generation of technical knowledge will be discussed here.

Nordhaus[14] mentions two characteristics of the production or generation of technical knowledge. In the first place, the cost of disseminating new ideas is generally small compared with the cost of generating technical knowledge. However, as technical knowledge becomes more specialized and specific, and its diffusion requires further new capital goods, the cost of diffusion can become high in absolute terms. A rapid diffusion of technical knowledge may also be hampered by the unavailability of sufficient trained personnel. In the second place, the generation of technical knowledge is accompanied by beneficial external effects,[15] since companies cannot reap all the benefits themselves. The acquisition of patents restricts these external effects. In the extreme case where companies are completely unable to profit from their new technical knowledge, the motivation for generating it vanishes.

Both points raise the question of how far the generation of technical knowledge yields public goods.[16] These are goods which cannot be split into smaller units for individual consumers, so that no consumer can be excluded from benefiting from them; the use of public goods by some consumers does not diminish the possibility of others using them—defence is often quoted as a typical example.

The acquistion of the rights to the use of technical knowledge by a firm excludes others from using that piece of technical knowledge but not from becoming acquainted with it. If the use of a piece of technical knowledge is kept within the firm, for example by taking out a patent, it is then possible to buy and sell the right of use on an individual basis in the market; if it is not kept within the firm, on the other hand, then the technical knowledge in question becomes a public good in so far as nobody is excluded from using it. It may therefore be said that the generation of new technical knowledge is something that gives rise to a potential public good, which in the case of complete disclosure becomes an actual public good.

This conclusion can be taken further if a distinction is made between devising new production methods and developing new products. Launching a new product entails a more explicit disclosure than the use of a new production method, so that the piece of technical knowledge that underlies a new product is more of a public good than that embodied in a new method of production. A distinction should therefore be made between the beneficial external effects of new production methods and corresponding ones derived from new products.

In an endogeneous microeconomic analysis of technical change the development of new production methods also opens up the possibility of choosing between labour-augmenting and capital-augmenting technical change. The studies dealing with this problem form the microeconomic complement to the macroeconomic considerations mentioned in Section 10.3. In this connection, Kamien and Schwartz[17] have determined how the level of investment in research and development affects the nature and rate of induced technical change, while Koizumi[18] has examined how the nature of technical change affects investment. In an important contribution to the micro-economic theory of endogenous technical change, Binswanger[19] starts from a description of the research process which is associated with expected pay-off functions in terms of efficiency improvements, and derives the invention possibilities. This study supports the criticism of Kennedy's model in Section 10.3.

After this brief discussion of the broader aspects of the generation and use of technical knowledge, we must find a criterion by which a company's contribution to technical development can be quantitatively measured. One frequently used is expenditure on research and development (R & D). This approach establishes a link with the sacrifice involved in the acquisition of new technical knowledge rather than with the results of the research work, so that it may lead to an overestimate of technical development. This overestimate increases if the R & D expenditure relates both to the acquisition of new technical ideas in the strict sense and to the work of refining the chosen technical possibilities into a form ready for production and marketing.

Mansfield[20] has carried out a thorough and extensive investigation into R & D expenditure, which, he found, rose by ten per cent a year in the USA in the period 1953–70. It is often said that research carried out in one sector is financed by another sector. In commercial companies the stress is not on basic research but on the development of new methods and products that are often known in advance to stand a good chance of being a commercial success. The profit motive therefore plays a predominant part in determining the nature and extent of R & D expenditure.[21] However, according to Mansfield,[22] the introduction of a new process or the launching of a new product requires more means of production than is apparent from the R & D expenditure. This last statement indicates that this expenditure is regarded not only as a measure of the expansion of technical knowledge, but also of the innovative activity of the firm that incurs it.

When the number of the firm's patents used by others is taken as a yardstick,[23] the generation and application of technical knowledge are measured not by the sacrifice involved but by the income resulting from these inventions. However, as we have seen before, it may be in the company's interest to keep some discoveries secret, and besides, not all the research workers' knowledge can be transformed into patents. Furthermore technical change involves not only spectacular inventions and innovations but also the stream of ordinary, minor improvements which generally require hardly any research facilities. Technical change and applications are therefore underestimated if measured by the number of patents, which in fact gives a better idea of the 'research input than of the research output'.[24]

Two other possible measures, based on the research output, are the profits from the extra production[25] and the number of important inventions. But the first is more a measure of the company's commercial activities than of its contribution to technical knowledge and the second involves too subjective an evaluation.

There is thus no single, fully satisfactory criterion whereby a company's contribution to technical change can be measured. A combination of R & D expenditure and the number of patents granted is potentially the most fruitful approach to measuring the advances of technical knowledge. This composite indicator, yet to be worked out in detail, can also serve as a measure of the innovative activity of the company.[26]

As mentioned before, the generation of technical knowledge yields a product that is potentially a very durable public good. To understand the production and use of this good a little better we must examine the market mechanism that regulates this type of production.

11.3 STATIC, DYNAMIC AND X-EFFICIENCY

With a given state of technology and perfect competition in all markets, the means of production are utilized to the full. This well-known statement about the optimum allocation of the factors of production implies that a maximum output is produced by the given factors of production, and at the same time the market and price mechanisms ensure that the use of the factors of production and the volume and composition of the output match consumer preferences.[27]

This type of economic efficiency is static in so far as one and the same state of technology is assumed. In fact, many economists, including Mansfield,[28] speak of a 'static economic efficiency'. However, there is also a dynamic interpretation of economic efficiency, in which the state of technology is not fixed and efficiency refers not only to the allocation of given resources but also to technical change. In developing the concept of dynamic efficiency, we shall concentrate on technical change in the narrow sense of the term. Our terminology will therefore differ from that of, for example, Scitovsky, for whom 'dynamic efficiency' refers to the rate at which a disequilibirum situation returns to equilibrium under the influence of market forces.[29]

If we take technical change in the narrow sense, i.e. an increase in technical knowledge, then dynamic efficiency includes the extent to which technology in the narrow sense is advanced. If, for example, we find that perfect competition offers little chance of doing research that may increase technical knowledge we must conclude that this market form is inefficient in the dynamic sense. However, dynamic efficiency can also be given a broad meaning, to include the applications of technical knowledge. This provides us with a framework to examine the significance of oligopoly and the size of the firm for technical change and its applications.

The treatment of the dynamic efficiency of oligopoly will not be restricted to technical change in the narrow sense because the technical possibilities reflected in the production function are not only determined by technical knowledge; new technical possibilities may also arise from using a better quality of labour, from the diffusion of existing technical knowledge and from changes in the size of the firm. In view of our terms of reference, furthermore, dynamic efficiency also encompasses technical change in the broad sense. If the potential and the realized technical possibilities are borne in mind, the dynamic efficiency of oligopoly should be judged by technical change in the narrow and the broad sense and by the actual advances of technology.

It has so far been assumed that production is technically efficient, so that only those combinations of the factors of production are taken into account which are involved in the production function. In the terms of the axiomatic production function, one is always at the boundary of the production set.

We shall now abandon this formulation of technical efficiency and examine those which take into account the fact that 'firms do not operate as efficiently as assumed in economic theory'.[30]

In an important step forward, Leibenstein[31] introduced the concept of X-efficiency, based on the realization that 'for a variety of reasons people and organizations normally work neither as hard nor as effectively as they could'.[32] He starts by assuming a connection between the degree of X-efficiency and the pressures of varying intensity exerted by competition. In a second article, he explains the existence of X-inefficiency by the fact that 'firms do not always introduce technical change when available and profitable'.[33] He develops his theory further by postulating areas of inertia, created by a certain resistance to change and faster work, which is noticeable at various levels within the enterprise. The profit motive becomes, within reason, secondary to the subjective advantages of an 'easy life'.[34] These areas of inertia also affect the diffusion[35] of technical knowledge. (McCain has recently put Leibenstein's theory into a broader context of cybernetics and information theory.[36]).

Working with Comanor, Leibenstein has used the concept of X-efficiency for calculating the adverse effects of monopoly on welfare,[37] called 'welfare losses'. Starting from the consideration that, owing to lack of competiton, X-inefficiency is relatively high in the case of monopoly, they conclude that such a system yields less welfare than a system characterized by free competition, partly because the allocation is not optimal and partly because the X-efficiency is low. It should be stressed that in this static welfare-economic treatment X-inefficiency is measured solely by the shortfall of production below the maximum level attainable with the given quality and quantity of the means of production. This also applies to Crew and Rowley's extension[38] of Comanor and Leibenstein's analysis, although this approach ignores the beneficial welfare effects of an easy life.

Williamson[39] has challenged Leibenstein's argument about X-efficiency, and Parish and Yew-Kwang Ng[40] have recently pointed out that it is not entirely clear to which phenomena the X-efficiency is relevant and to which it is not. Even our brief summary of the literature shows that in some cases X-inefficiency is an inefficient use of the available means of production, while in other cases it refers to a failure to utilize all the possibilities that technical advances offer, for example because of inefficient management. What does invariably emerge, however, is that X-innefficiency is due to internal factors.[41]

It seems sensible to distinguish between X-efficiency in the static and dynamic sense. In the former case production is characterized by X-efficiency if it gives the maximum output attainable with the available quantity and quality of factors of production. X-efficiency then boils down to technical efficiency and is based on a given state of technology, expressed by a production function. It is only X-efficiency in this static sense that Leibenstein and others have introduced into the welfare-economic treatment of monopoly. X-efficiency in the dynamic sense means that the firm is making use of one or more up-to-date technical possibilities. X-inefficiency in this sense occurs when techniques are used which are obsolete from the point of view of cost. The different ways of evaluating existing capital goods, based on the depreciation method used, of course complicate the assessment of this dynamic X-inefficiency. A firm without any X-inefficiency in the static sense may still exhibit X-inefficiency in the dynamic sense if its combination of the factors of production belongs to a production function that reflects the state of technology at an earlier point in time. Using the axiomatic production function, this means that while a point on the boundary of the production set is chosen the production set itself is obsolete. In a dynamic context X-efficiency is no longer purely technical efficiency, because the delay in effecting technical advances can, for example, result from an outmoded organizational structure, the use of a certain depreciation method or inefficient managers. Past decisions about the use of scarce resources will then also affect the firm's X-efficiency in the dynamic sense. This is consistent with the conclusion in Section 8.7 about the nature of the production function in a dynamic context.

We must now examine the connection between the static and dynamic X-efficiency of the firm's production methods and the interpretation of dynamic efficiency developed before. X-efficiency in the static sense complements dynamic efficiency, since the latter concerns the extent to which technology is advanced and applied. The firm can raise its output by producing more with the same means of production, so that a higher static X-efficiency is reached. Growth may also result from the development and application of new techniques, and this affects the dynamic efficiency of the company. If the latest techniques are indeed employed, i.e. if the most recent production function applies, we can also speak of an X-efficiency in the dynamic sense. However, the greater the use of older, more expensive techniques by the firm, the higher is the X-inefficiency in the dynamic sense and also the lower the dynamic efficiency. Thus the concepts merge into each other with the blurring of the distinction between technical and economic efficiency in a dynamic context. Yet the distinction remains useful even here, for example in describing the case where a firm employs a new technique due to market pressures even though cost does not justify this. Ignoring the new method would not mean an X-inefficiency in the dynamic sense, but it would lower the dynamic efficiency. Furthermore, the extent to which new technical knowledge is generated is also important when assessing the dynamic efficiency of an enterprise or an industry.

These concepts also apply when technical change is treated as an endogenous factor and is regarded as a learning process; the dynamic efficiency of a firm here refers to the application of techniques whose productivity increases as more and more experience is accumulated. But the more production falls behind, despite the learning process, because of the use of obsolete and relatively unprofitable techniques, the greater the X-inefficiency in the dynamic sense.

To sum up, the internal affairs of the firm and the role played in them by mananagement are the main factors relevant to X-efficiency in the static and dynamic sense, whereas the firm's position in the market is what is primarily relevant to dynamic efficiency. There is not, however, simple duality since the internal and external affairs of an enterprise are closely intertwined.[42]

11.4 TECHNICAL CHANGE AND THE SIZE OF THE FIRM

At a given state of technology and overall size of the market, the number of firms operating in an industry depends on the economies of scale from large-scale production and to what extent these economies are utilized. Where these economies, which are predominantly technical, peter out beyond a certain level of production there is in principle an upper limit to the number of firms; the lower limit is set by the level of production below which there are only diseconomies of scale. Only in the special case where the economies of scale immediately and permanently change into diseconomies do these two limits coincide and one can speak of an optimum size of the firm.[43]

If we now assume that the overall size of the market is not fixed but varies, for example due to the commercial policy of the firms, then the scale effects resulting from the elements of the marketing mix may show a different pattern from the technical scale effects, so that the upper and lower limits of the number of firms in an industry will change as a result. If the overall market is growing and the available economies of scale are always utilized, the technical and commercial scale effects influence the limits within which the number of firms in the industry will fluctuate.

Companies may have more than one plant, so that a new source of scale effects appears as a company grows in size; in particular, the management of a concern can extend over several plants or operating units in an industry. Although the usefulness of this is as subject to limits as any other given amount of a factor of production, it often produces economies of scale before these limits are reached. For this reason, both the upper and the lower limit of the number of firms in an industry are disproportionately lower than the corresponding limits of the number of plants. Consequently, while it is necessary to distinguish between the horizontal concentration of firms and that of plants (including the case where the firm has plants operating in different industries), scale effects can arise from the wide span of control and affect the vertical and conglomerate concentration[44] of the firms. If beneficial scale effects of this kind are utilized, so that the company grows both internally and externally, the firm concentration is disproportionately greater than in the case where the plants operate as independent units.

The complicated picture of the effect of technical change on the optimum size of the firm was discussed by Young in an article published as long ago as 1928, which was for long ignored despite its sound comments.[45] While the static allocation theory assumes either constant or diminishing returns to scale, Young believes that technical improvements are in fact an important cause of increasing returns to scale. The utilization of these for increasing production in turn promotes technical change, so that an endogenous and self-reinforcing process is in operation. In Kaldor's words, economic theory pays too much attention to the market mechanism as an instrument for allocating the factors of production, and it pays too little attention to the creative effect of the market with respect to economic growth. According to him, the whole concept of equilibrium in economic theory is closely connected with allocation at a given state of technology. He views the interactions between commercial expansion and technical change as an important aspect of economic growth in which equilibrium is absent, owing to increasing returns to scale. 'Change therefore exerts constant pressure for further change.'[46]

Hahn[47] has criticized Kaldor's views on equilibrium in economics. He believes that one can speak of efficient allocation of resources even in the case of increasing returns and that the greater the potential losses through misallocation the more important the increasing returns. He overlooks the fact that Kaldor did not deny the importance of the market for the allocation of resources, but laid more stress on other aspects of the market mechanism which are closely related to technical development as a dynamic force. In Nelson and Winter's words: 'the diversity and change that are suppressed by aggregation, maximization and equilibrium are not the epiphenomena of technical advance. They are central phenomena.'[48] Empirical research by Viaciago[49] and La Tourette[50] indicates that increasing returns are less important in fast-growing countries than Kaldor suggests.[51]

To evaluate this view of the economic process, it is useful to distinguish between the potential scale effects from the acquisition and diffusion of technical knowledge and the scale effects from the application of new techniques. Important advantages of scale can be obtained by applying the same technical knowledge in a number of companies, in a number of plants within the same company, and for a number of products within the same company. An enterprise that supplies various markets with a wide range of products has a larger base for the use of new inventions than one that makes only one product.

The variety of production methods which is characteristic of a multi-product company increases the number of points at which innovations based on the same technical knowledge can be applied. The sale of new technical knowledge by incorporating it in new production methods and new products yields more and more economies of scale, which constitute a major incentive for internal growth by diversification. A quite different alternative available to the management is to broaden the platform for technical knowledge and intensify cooperation with other firms, or even merge with them.[52] This accelerates and simplifies the diffusion of technical knowledge, as can be seen in the case of multinational companies.[53]

The scale effects expected from the actual introduction of new techniques are less clear-cut and depend to a great extent on the nature of these techniques; in general, a predominantly labour-saving technique increases the optimum size of the firm, while a predominantly capital-saving one is expected to reduce it; a more definite statement can be made only if we know more about the nature of the new techniques.[54] An invention whose application leads to increasing returns to scale and thus increases the optimum size of the firm can greatly promote the internal growth of the firm, provided that the market can absorb the extra output. With expanding productive and commercial activities, the monopoly power of the firm generally increases.

Technical development does not in itself ensure a sound management and the resulting expansion of the firm, for the utilization of technical change depends on company policy and the financial situation. Even large firms can get into considerable difficulties if they operate with a high static and dynamic X-inefficiency. Owing to the very high demands placed on management in our dynamic society, the line between demonstrable human error and unavoidable mistakes is becoming more and more difficult to determine, notably in the case of decisions about the optimum size of the firm.

Economists have put forward various arguments to show that the larger the firm, the more it promotes technical development. Besides the high cost of modern research,[55] they often mention the great financial risks involved. Large firms can weather unsuccessful projects and ventures better than small ones and can profit from the economies of scale offered by larger laboratories. Conglomerates and multinational concerns possess considerable scope for the application of a great variety of new production methods, as well as for the marketing of new products. With technical change in the broader sense, greater division of labour is realizable within a large company and there are greater facilities for internal training.

A counter-argument is that these large laboratories are in many cases bureaucratic, suppressing the creativity needed for inventing and developing new methods and products, and delaying their implementation. Sometimes, because of the large size of the firm, innovations made in the laboratory are deliberately withheld. A final point is that many innovations do not in fact come out of research laboratories.

Nelson's hypothesis[56] that new technical knowledge is mainly generated in firms that make a great variety of products has been tested, for example, by Comanor[57], Grabowski[58] and Wood[59].   Although Grabowski's findings lend some support to this hypothesis, econometric investigations in general do not confirm it. These negative results do not however, disprove the more general hypothesis that larger firms spend relatively more on research and development than small ones, since a company can grow even if its range of products remains the same. In the attempts to test this hypothesis by empirical investigations, an important point is whether the intensity of the research activities increases with the size of the firm, i.e. whether large firms spend relatively more on research. Research intensity is measured by R & D expenditure as a percentage of turnover, or by research staff as a percentage of total employees. In extensive investigations covering seventeen branches of industry, Hamberg[60] has found that the number of R & D people employed, but not the research intensity, increases with the size of the firm; indeed there are signs that, beyond a certain size, R & D expenditure and the number of patents taken out decreases relatively.[61] Recently empirical investigations in Britain found that 'firms of all sizes' take part in inventive work, as measured by the number of patents granted.[62] Moreover, it cannot always be established that the largest firms are the quickest in applying new techniques, though Mansfield initially amassed a great deal of statistical data pointing to this;[63] later, however, he wrote that 'there is little evidence that industrial giants are needed in most industries to ensure rapid technological change and rapid utilization of new techniques'.[64] Similarly, Adams and Dirlam, in their well-known study of the steel industry, did not find that large firms are quicker in introducing new techniques than small firms.[65] In an exchange of views with McAdams,[66] they remark that 'it is the cold wind of competition and not industrial concentration which is conducive to innovations and economic progress.'[67] On the other hand, a study by Branch[68] supports the hypothesis that research activity tends to increase profits and intensify growth at the same time. It has also been pointed out that firms active in sectors of advanced technology undertake R & D activities at a rate higher than a certain minimum (the threshold)[69] in order to stay in business.

On the basis of theoretical analysis and empirical research, we can put forward the following hypothesis, amenable to testing and possible falsification, about the connection between technical change and the size of the firm. Owing to economies of scale which usually follow from the generation and diffusion of technical knowledge the contribution to technical development at first increases more than proportionately with the size of the firm, but beyond a certain size research intensity remains, at best, constant, since the possibilities of applying technical knowledge are limited and there are diseconomies of scale attached to research. In many cases, research activity may therefore diminish beyond a certain firm size.[70] There are physical, organizational, financial and commercial limits in the firm imposed on the application of new production methods and on the introduction of new products. Since the scale effects from the application of new inventions can be either advantageous or disadvantageous, the relatively quicker application of new techniques by large firms can come to an end sooner than does the increase in their research intensity. The relative position of the two change-over points is partly determined by the power relationships in the market.

This hypothesis ultimately leads to the conclusion that the number of firms supplying goods in a growing market is limited notably by the scale effects of technical knowledge. On the other hand, the application of new techniques produces such disadvantageous scale effects beyond a certain firm size that generally no complete monopoly positions will arise. Thus the nature of technical change to some extent explains why oligopoly is so common.

11.5 OLIGOPOLY

Oligopoly is generally regarded as a market form or type of competition in which there only a few suppliers and a large number of consumers.[71] In practice, there are almost always significant differences between the goods offered by the different manufacturers. This product differentiation makes for a heterogeneous oligopoly, and we refer below to a type of oligopoly whose heterogeneous character is due to differences in the location of the companies, the quality of their products, the services they offer, their packaging and their advertising. Each supplier is, as it were, a monopolist in his own goods, which compete with related goods offered by other suppliers. Oligopoly does not necessarily mean that the companies involved are large; just as we have small monopolists, there are also small oligopolists.

To understand the nature of oligopolistic competition, a distinction must be made between market form and market behaviour. In the case of oligopoly, the market form permits a variety of behaviours, which means that no specific outcome can be predicted from the market process. A special complication is that the market behaviour may be directed at changing the market form, and may in fact achieve this, for example when some oligopolists decide to merge. Owing to the great variety of oligopolists' behaviour patterns, the theory of oligopoly has a marked heterogeneous character.

The question of prices has long been the central issue in the theory of oligopoly. One of the most difficult problems in the whole of economics is to explain the level of prices in a heterogeneous oligopoly, largely because of the interdependence of oligopolists. If the sales of one oligopolist are regarded as a function of his price and of the prices of his competitors, then his profits depend on all the prices involved. Individual profit maximization then requires a more accurate specification (reaction hypothesis) of competitors' behaviour.

The heterogeneous character of the theory of oligopoly is partly due to the possible variations of the reaction hypothesis, since changing the assumption about the competitors' reaction to a price change of one oligopolist will change the set of equilibrium prices.

The theory of oligopoly is not restricted to individual profit maximization, and Fellner has examined in detail the idea of the maximization of joint profits. His theory can be regarded as an extension of Chamberlin's views on oligopoly.[72] Baumol has developed another variation, in which, given a certain profit level, the main aim of management is to maximize turnover.

Generally speaking, two assumptions are encountered in the price theory of oligopoly: that oligopolists make a collective decision, or that they all make individual ones. In the first case, a cooperative behaviour pattern is proposed, and the equilibrium position is Pareto-optimal from the oligopolists' point of view.[73] In the second case, a non-cooperative behaviour pattern is assumed.[74] It has also been suggested that the oligopolists sometimes display cooperative and sometimes un-cooperative behaviour, either in connection with pricing or other components of the marketing mix. This approach has been chosen by Williamson for his interesting, though little noticed, model of oligopoly.[75] In the theory of games, too, both behaviour patterns are taken into account.

We get a one-sided picture of oligopoly from prices. The oligopolistic entrepreneur can also vary the quality of the goods supplied, advertising, and the services offered. Oligopolistic entrepreneurs do generally seem to put more emphasis on non-price than on price competition. Heflebower[76] has pointed out that, at least in the short run, non-price competiton does not cause a serious disturbance in the market. If the price status quo is generally being maintained, competition can be reduced by a formal cartel agreement, but this may not be necessary where there is a tacit understanding. For this purpose one oligopolist is sometimes accepted as a 'price leader', although price leadership may also come about without any joint profit maximization. Or there may be agreements about mutual research.

As we look at oligopolists' non-price policy, the assumption that management only aims at maximizing profit becomes more dubious, for—as Kotler rightly remarks—'the company generally pursues several goals simultaneously'.[77] With the introduction of a range of products entrepreneurs seem to put a higher premium on continuity and a steady increase in the trading results rather than on immediate profit maximization.

We must also examine the fact that existing firms take into account the possibility of other firms entering their industry, for example because of the relatively high profits. They are faced with the dilemma of whether to try to stop this (by price policy or other means), or not to, with the eventual possibility of a smaller profit margin. In a dynamic context, it is precisely by selling at a relatively low price that the existing companies may encourage the entry of new competitors into their field, owing to the resulting great increase in demand and the limited possibilities of financing innovations from profit. When newcomers can enter the market easily, the established firms realize their optimum size sooner than they do when it is almost impossible for newcomers to join.[78] This theory deals in detail with the factors that prevent entry, such as the absolute cost advantages of the established firms, product differentiation, the available technical knowledge and large-scale production.[79] Some economists have also listed marketing cost as a barrier to entry.[80]

It has long been assumed in the theory of oligopoly that oligopolistic firms are roughly of the same size ('symmetrical oligopoly'), but increasing interest has recently been shown in the great variety of the absolute and the relative sizes of these firms. One or a handful of oligopolistic firms can assume a dominant position in the market by virtue of their large relative size ('asymmetric oligopoly').[81] A high degree of concentration, furthermore, produces a 'tight' oligopoly, while a low one produces a 'loose' oligopoly.

All this adds up to a variety of oligopolistic market forms and behaviour. The oligopolistic firms can be small or large in absolute terms, differ slightly or considerably in terms of relative size, form part of larger concerns or not (which may or may not extend over a number of other markets), come to a formal agreement with one another or not, compete with one another to the bitter end or establish a peaceful coexistence in certain areas; they can behave in a cooperative or non-cooperative manner; and when they decide to block the entry of would-be competitors into the market, they can do this in many different ways.

Almost any statement about oligopoly is partly true, while none is entirely true. Oligopoly combines monopolistic and competitive elements in an unpredictable way that varies according to time, place and circumstance, it leads to a market performance that cannot be rigidly forecast from market conditions and participants' behaviour patterns alone. The varied character of oligopoly thus foreshadows the qualified judgements that emerge below from the analysis of the significance of this market form for technical change.

11.6 OLIGOPOLY AND TECHNICAL CHANGE

In a penetrating study on whether monopoly helps or hinders technical change, Hennipman comes to the conclusion that 'simple and sweeping generalizations, still frequently encountered, have no foundation either in theory or in fact'.[82] However, some economists still propound quite general statements about the dynamic efficiency of oligopoly and about its dynamic X-efficiency. In their opinion oligopoly as a blend of monopoly and competition combines the effects of these two market forms on technical change.[83] It is so risky and costly to develop and apply new production methods that a certain market dominance, coupled with financial power is necessary. Active competition from existing companies and the threat of new competition constantly forces oligopolists to improve their production methods and update their products.

This view of the dynamic efficiency of oligopoly is based to a large extent on the assumption that 'administered prices' prevail, an administered price being—according to Means, who coined the term—'a price set by someone, usually a producer or seller, and kept constant for a period of time and for a series of transactions'.[84] Administered prices introduce the possibility of price rigidity. Over a considerable period of time, changes in the oligopolists' costs and sales do not lead to changes in prices that correspond to changes in the microeconomic data.[85] Particularly important here is the fact that in most countries the rise in the productivity of oligopolistic enterprises leads not to a fall in prices but to a rise in wages under the influence of collective bargaining, and this rise in wages stimulates the search for new production methods.[86] Being administered, prices are not dictated directly by the market, but are instead decided in the framework of overall company policy. Price policy is then aimed mainly at securing sufficient profits to finance technical development. New production methods are developed and applied in order to improve the internal cost situation, while existing products are improved and new ones are offered in order to maintain and improve the position on the product market. In this way, technical change and the oligopolists' commercial activities are intertwined in a way which reflects the fact that administered prices are part of overall company policy.[87]

Administered prices may form the cornerstone of oligopolists' dynamic behaviour, but this is by no means necessary. If for example a price cartel prevents potential competitors entering the field, and if the existing companies display cooperative behaviour, they may easily adopt a conservative attitude both towards the development and application of new techniques, and towards the launching of new products. In a variant of this situation, a research cartel is set up alongside the price cartel, so that technical knowledge is freely pooled but restrictions are placed on its application, similar to those already applied to prices.

Once a research cartel has been formed, the important point is whether the new technical knowledge generated inside the oligopolistic company is useful for maintaining the position of the latter in the market, and whether it can be applied by the company for its own good at the time it chooses. The less the individual power positions are threatened by the market behaviour of the existing companies, the weaker the company's urge to find new production methods and new products. A rapid diffusion of new technical knowledge, for example through rapid staff turnover, weakens the urge to invest heavily in research and development. On the other hand, the relatively great monopoly power guarantees monopoly profits for the oligopolists in this more or less protected situation, and the financial latitude thus obtained will stimulate research.

In such a tight oligopoly, research may sometimes by preference be directed at the development of new products. Most goods have a natural lifetime, to some extent independent of the market behaviour of producers towards their products. When a product enters the last stage of its life, the oligopolist finds it advisable to start looking for a replacement. In this way, intense innovative competition over products can spontaneously arise, while new production methods receive less attention. However, much depends on the many possible ways oligopolists react. In a recent study Kamien and Schwartz suggest the possibility of an optimal degree of technological rivalry.[88]

The threat of new competitors can militate against the tightening of oligopoly. This entry can be considerably hindered by maintaining and increasing the lead in technical knowledge, since one can assume that newcomers would find it considerably more difficult to appreciate the newly acquired technical knowledge of a firm than the older established firms. Investment in the generation of technical knowledge is therefore more use in combating potential competitors than in competing with existing ones. If the established firms want to prevent a loosening of the oligopoly, they can adopt cooperative behaviour towards the updating of products and production methods, just as they can towards prices. In this case, technology is promoted to a smaller extent than in the situation where there is non-cooperative behaviour.

The application of new technical knowledge in oligopoly can thus be equated first and foremost with the introduction of new production methods. If a new technique is embodied in new capital goods, the existing capital stock will obstruct its introduction, particularly if some of the machines have only recently been installed. In the absence of external pressure, a delay in introducing the new technique can even make for X-inefficiency in the dynamic sense.[89] However, an accelerated introduction of new techniques, under the influence of actual and potential competition, is equally possible.[90] Kamien and Schwartz have compared these behaviour patterns with the timing of innovations.[91]

The question of optimum timing also arises in the case of the launching of new products. Economists often maintain that it is in the monopolist's interest to postpone the launching of a new product. It is important to clarify whether one has actual or potential competition in mind.[92] Established firms may see little sense in launching a product sooner than necessary if the existing product range is sufficiently profitable. However, the threat of entry can speed up the introduction of new products, since expectations about future profits are then changed.

The position of established firms with respect to one another partly depends on the degree of concentration in the oligopolistic market. The choice of the time for launching a new product is affected by monopoly power more in a tight oligopoly than in a loose one (there are of course numerous intermediate cases). The policy of some oligopolists to extend their range of products regularly can be regarded as one of the concrete results of the balance of forces between their power in the market and the internal affairs of the firm. However, it is also possible that in the case of asymmetric oligopoly the smaller firms' urge to update and the larger firms' facilities for it result in the former concentrating on a few special products, and the latter widening their product range.

In view of the variety of oligopolistic market structures, it is not surprising that the econometric studies of the connection between oligopoly and technical change in fact refer only to certain cases of oligopoly. The oligopolistic market position is generally identified with a large company, and a symmetrical, tight oligopoly assumed. After correction has been made for the differences between the various industries' available technical knowledge, Horowitz's[93] and Scherer's[94] investigations indicate a very weak connection between the research intensity and the degree of horizontal concentration. There is also a difficult problem concerning the direction of causality: how far is market concentration a result of the research intensity, particularly if there is a connection between the latter and the possibilities of increasing technical knowledge?

The few empirical studies that examine the effect of market structure on technical change suggest that the latter depends less on monopoly power than might be expected from the facilities of companies with considerable monopoly power.[95] However, great care is needed when one tries to draw general conclusions from this. In the first place, some of the technical advances in companies are made outside the research budget and without the involvement of laboratory staff. The continuous stream of improvements in production methods brought about by ordinary technical people is relatively wider in an oligopolistic company than in a completely monopolistic or perfectly competitive company. Furthermore, a symmetrical, loose oligopoly may exhibit a stronger connection between the degree of concentration and R & D expenditure than does a symmetrical, tight oligopoly. In the case of asymmetric oligopoly, any hypothesis should distinguish between small and large enterprises. In addition, the role played by the threat of newcomers' entry is important. There is not yet much empirical work[96] on this topic, but it seems obvious that the gap between the *ex ante* possibilities and the actual *ex post* situation is greater when the market is protected against such entry than where the fear of entry urges established companies to utilize their potential to the full. This is supported by Carlson's investigations of the efficiency of Swedish industry in 1968, which show that those branches of industry which are sheltered from foreign competition are less X-efficient than those which are not.[97]

Moreover, Tilton[98] has pointed out that most investigations are aimed at establishing a connection between the characteristics of the firm and concentrate on just one part of what he calls the technological development cycle, although it is the whole of this cycle that is relevant. The company that makes the greatest contribution to the development of new technical knowledge need not be the one that rapidly applies and diffuses new ideas.[99]

Finally, there is the inevitable difficulty of placing the individual research worker's inventiveness in proper perspective, even when one restricts this to a historical account of how inventions have come about.[100] The more the inventor is overshadowed by R & D expenditure, the size of the laboratory, the size of the firm, the market situation and the commercial policy of the company, the more difficult it is to discover from statistical material the origin of new technical knowledge.

Microeconomic empirical studies of the connection between the size of the firm, the market form and technical change are therefore investigations into the prerequisites for technical change rather than inquiries into its causes. Since the human mind is particularly unpredictable when it comes to inventiveness and creativity, it is not surprising that these investigations give ambiguous answers.

This does not mean that the empirical results are unimportant. If there are complications in the statistical investigations, caused by the definition of concepts such as the size of the firm, the yardstick for measuring the degree of concentration,[101] monopoly power and R & D expenditure, it is the task of theoretical economists to make the analysis more incisive and precise. A good example of the interaction between empirical and theoretical work is provided by the attempts to separate quantitatively the effect of economies of scale from the effect of technical change on the volume of production.[102]

On the basis of both theoretical hypotheses and the results of empirical research, we can come to the following conclusions. As the monopoly power of a company increases, so does its ability to increase technical knowledge and to apply it in the form of new production methods and new products, but at the same time its urge to update its technical knowledge and its inclination to introduce innovations decrease. The looser the oligopoly and the less protected it is against the entry of new competitors, the more the company's actual contribution to technical development will approximate its potential maximum.

This hypothesis implies that a monopolistic firm that need not fear the entry of new competitors will contribute less to technical development, relative to its size, than a firm which operates under conditions of perfect competition, in a market not protected from such entry. The more the oligopolistic market structure approximates to one or the other of these extremes, the more the innovative performance conforms to the corresponding extreme.[103] Since innovation is rather modest at both extremes, though for different reasons, we may conclude that innovative market performance is likely to be better in the various intermediate forms of oligopoly than in either monopoly or perfect competition.

11.7 THE LABOUR MARKET

We have already discussed, in Section 10.4, the macroeconomic aspects of the relationship between technical change and the labour market, and concluded that technical change is not generally expected to cause any considerable or permanent unemployment.

Now that we have analyzed the microeconomic aspects of technical change, it will be interesting to re-examine the situation of the labour market. The most striking conclusion here is that a large and permanent reshuffling can be expected between the various segments of the labour market.[104] Not only is continuous retraining necessary, but redundancies will occur whenever the application of new techniques renders part of the labour force superfluous. The disappearance of older products and the introduction of new ones can cause drastic changes in the pattern of employment; the application of a new technique may be hampered by the unavailability of appropriately trained labour, as a result of which some other people cannot find jobs either. The importance of full employment for welfare should never be underestimated. Some rash statements have been made, for example by those who believe that large-scale, short-term unemployment is necessary for the good of the environment. We see at every turn how important having a job is from a social and psychological point of view, and how much personal suffering people go through when they lose their jobs, for whatever reason. Serious problems of individual adjustment also arise when, owing to technical development, firms change over to larger production units, either by internal growth or by external expansion. Many people are alienated from the new work situation, and there may be important psychological problems, aggravated by uncertainty. When a firm changes over to completely new methods of production, which usually entails a reshuffling of personnel, psychological counselling is often necessary, to avoid the sense of security of some employees being undermined.[105]

Large oligopolistic firms sometimes exhibit a tendency to treat labour as if it were an end-product. This tendency is prompted by the continual changes facing the management of such companies, and by the undervaluing of the beneficial welfare effects of the sheer opportunity to work *per se*. We must hope that decisions about the choice of new production methods and new products will in future be made in such a way that the characteristics of the available techniques, on the one hand, and the number and type of employees involved, on the other, will be considered in a more balanced manner.

11.8 CONCLUSIONS

Technical development stems from individual human activities, so that a study of its microeconomic background is indispensable. Oligopoly exhibits not only different behaviour, reactions and expectations of oligopolists, but also different methods of updating technical knowledge. The power usually wielded by oligopolistic companies in the market does not mean that technical development within the firm can necessarily be channelled in the direction preferred by the management.[106] Creativity can be organized within certain limits, but it does not take orders. Oligopoly can combine dynamic efficiency with dynamic X-inefficiency as regards the application of production methods. We can derive from this an argument for finding 'policies which will act directly upon managers rather than through the market'.[107]

The likely future trend will be to base decisions about technical change not on simple criteria like financial yield, but on more basic, subjective preferences about the nature, direction and rate of technical development and its applications. This is already happening with the position of labour in some firms undergoing rapid and drastic technical change.

The less employees regard technical change as an inevitable fact, the more they will want to direct it and at least mould it to suit their own purposes. This will influence the rise of real incomes, as well as income distribution. A temporary opposition to the introduction of new techniques, for example to save some jobs, usually causes a drop in the firm's profits, so that not infrequently the income of others in the firm will also be affected. This in itself places high demands on the decision-making apparatus, and these demands are increased by the international aspects of competition in the case of many firms. Technical change may also be deliberately hindered on the grounds that the external disadvantages of an increase in production outweigh the advantages of the application of new techniques. Sylos-Labini[108] has done some important theoretical and empirical work on the connection between productivity, inflation and the behaviour of trade unions, and concluded that productivity is independent of wages in the short run, while in the long run the firms that cannot pay higher wages simply disappear from the scene and the remaining firms try to improve productivity. This implies that, in the long run, trade-union action is not neutral with respect to growth, income distribution and productivity.

Codifying the various microeconomic behaviour patterns into new theoretical concepts will definitely change the predominantly macroeconomic views about technical change.[109]

Very difficult problems are raised by the effect of technical change on the size of the firm and on the concentration of companies and plants in the market, and conversely by the effect of these on the nature and rate of technical change. These relationships concerning market power will also influence the distribution of incomes and assets in society, not to mention economic and political power. We now turn therefore to an examination of the relationship between technical change and the economic system.

12 Technical development and economic policy

Difficult problems arise when economics leaves the serene field of pure science and enters economic policy where deliberately or otherwise it comes under the influence of non-scientific forces, ideologies, interests and emotions, amidst which it has to hold its own and define its own specific responsibilities.

Hennipman[1]

12.1 INTRODUCTION

The nature and rate of technical change are not independent of the economic system. In an economy with strong central control, for example, many aspects of both the generation of new technical knowledge and the application of technical inventions are closely controlled and programmed by the government.[2] As a result technical advances in these societies reflect the preferences of the central government. In contrast, in an economy based largely on the operation of the price mechanism, technical change is closely linked with market forces which involve widely divergent preferences as regards the direction of research and innovation. (Studying capital-labour substitution and technical change in planned and market-oriented economic systems, Asher and Kumar[3] have found that, between 1950 and 1970, the annual rise in the productivity of the manufacturing sector was 0.15–2.86 per cent higher in the USA than in the USSR, which almost certainly reflects a wider spectrum of preferences in the former).

Conversely, economic systems are partly the result of past technical change. Technical change affects both the ends and the means of economic policy. As technical knowledge grows both qualitatively and quantitatively in a planned economy, decentralization of decision-making sooner or later becomes inevitable, if technical development is not to come to a standstill.[4] The central body lacks both the inclination and the ability to recognize the creativity of technical people, and even the comparision of various production methods calls for specialized technical knowledge. After some time, the heterogeneity of production methods is followed by a heterogeneity of goods to which consumers' preferences are directed. Owing to technical change, elements of a free-market economy begin to complement the initially centralized system, although political factors may impede this process for some time. Conversely, technical development in an economic system based mainly on the price mechanism leads—for example because of external effects—to a partial centralization of decisions about the direction and rate of technical change. It is open to discussion whether these elements of free-market economy in one case and of planned economy in the other result from technical development as such (an objective factor) or from people's behaviour patterns (a subjective factor), but in either case economic systems are not in reality independent of technical change.

One may therefore expect that, under the influence of technical development, planned economies and free-market economies will converge or approximate to each other, depending on the political decision-making processes. The actual convergence[5] is the resultant of more or less mechanically determined objective factors and of subjective factors based on the behaviour of individuals and groups.

This convergence between the Eastern and the Western economic systems should be clearly distinguished from Tinbergen's convergence theory, in which convergence, determined exclusively by objective circumstances, follows an inevitable course. This inevitability rests particularly on Tinbergen's notion that an optimum economic system, lying somewhere between centralization and decentralization, can be obtained in a positive rather than a normative manner.[6] Van den Doel has developed Tinbergen's theory further by assuming that every economic system evolves towards its own optimum.[7] This relative version of the convergence theory gives the impression that an optimum can be specified for an economic system independent of the preferences of the economic subjects. If these preferences are taken into account, as many optimum economic systems can be devised as there are people. A second objection to the theories of Tinbergen and van den Doel is that they are mostly static. If there is indeed one optimum economic system why was it not reached some time ago? In fact, convergence is connected with the dynamics of technical change, which is influenced by human affairs and preferences.

We again confine the discussion mainly to the economic system of what is known as the West,[7] bearing in mind that technical development in principle continuously changes the economic system because its welfare effects influence economic policy. We have seen in the previous chapter that technical change affects both the optimum size and the monopoly power of the firm. The power relationships in the market can be profoundly changed by an increase in technical knowledge, the application of new techniques and the introduction of new products. The aim of a policy about competition may be either to promote or to hinder the various forms of concentration, but technical development can alter the acceptability of certain types of competition.

The state has long been interested in certain aspects of technical development. In the era of mercantilism it opposed or prohibited the export of machines on the grounds that it would help other countries to build up a competitive industry, and the attraction of trained people from abroad was also part of government state policy at that time.[8] The age-old policy on patents, according to which inventors are granted exclusive rights for exploiting their inventions on a commercial basis for a certain number of years, has been of great importance,[9] for it has stimulated inventions and the propagation of knowledge, and, as a result, moved scarce resources into research. This welfare effect must be weighed against the creation of a monopolistic position, which can serve as a starting point for further monopolization even after the life of the patent has expired. 'To buy innovation by paying with unnecessarily long delays of imitation is a poor bargain for society to make', says Machlup.[10] We shall return later to the normative assessment of these welfare effects.

Another form of protection is the temporary imposition of import duties on foreign goods. The 'nurturing' argument used here is mostly associated with List, who has in mind particularly infant industries where important technical innovations are being developed.[11] The justification of this violation of free trade is based on the consideration that the learning process in the protected enterprise or industry in question leads to external advantages, connected for the most part with the diffusion of technical knowledge among other firms within the national economy.[12]

If the formal and subjective concept of welfare is adopted, other welfare effects can also be brought into the picture. Particularly important are the welfare effects of satisfying subjective needs by the initiation and promotion of technical development. In a dynamic approach, not only current needs but also those of future generations are taken into account.

Besides giving rise to interesting welfare ramifications (discussed below), List's ideas have played an important role in the dynamic theory of international trade, whose exponents start with the life-cycle of products and explain the changes in the pattern of international trade by the lead a country acquires by being the first to exploit an invention.[13] The production and export of a good follow the well-known sigmoid, or S-shaped, path of the life cycle, the falling section of which is closely connected with production in the imitating countries. Just as there is the problem of the optimum period for import duties in the nurturing argument, what period should, ideally, elapse between innovation and imitation in a world-wide context?[14] The establishment of a connection between international trade and technical development manifested in new products could, therefore, be of great help to the theory of international trade.

In their excellent and famous review articles published in the mid-1960s, Bhagwati[15] and Chipman[16] scarcely mentioned technical change in their explanation of international trade. (Johnson's studies deserve special mention here.[17]) Although we still await a comprehensive work on the subject, more recent publications[18] and empirical investigations[19] have devoted more time to the connection between technical development and international trade. In this field economists have mainly dealt with the application of new techniques, but international trade in technical knowledge may also be used as a starting point for an economic analysis, as, for example, in Rodriguez's pioneering work.[20]

The question of how technical development should be

stimulated—if at all—in the Third World within the framework of a global development strategy and the international division of labour is of enormous practical and political importance and has been part of the discussion about the nature, direction and diffusion of the growth process in the developing countries.[21]

Tinbergen has called for a world-wide plan for the international division of labour, which involves recommendations on the establishment and transfer of various branches of industry.[22] As can be seen from Chakravarty's excellent study,[23] great advances have recently been made in the field of optimum programmes for economic development, but the choice of investment criteria still involves conceptual difficulties of a political nature. Bruton[24] came to the same conclusion after analysing the pros and cons of expanding industry or agriculture in the developing countries. Even if a decision between labour-intensive and capital-intensive production methods is ostensibly based not only on the immediate effects on the growth and composition of the national product but also on future requirements, the learning process involved in advanced techniques and the importance of employment for the individual, the choice is of only formal significance because of the intrinsically subjective nature of the welfare effects involved.[25] The problem is further complicated—if that is possible—by trying to introduce into the analysis of optimal technical development policy, the magnitude, nature and trend of the research expenditure. However, these complications do not invalidate Tinbergen's method in which ends and means are defined and, as far as possible, quantified,[26] to help decision-making. In a very useful study, Sen[27] deals with the institutional features, political feasibility and behavioural characteristics that in an underdeveloped economy influence the technical possibilities of improving employment. We need to assess technical change from a welfare-economic angle, irrespective of whether it is the technical change of a firm, a sector, an economy, a group of economies or the world economy. We cannot restrict ourselves here to static welfare theory, since the preferences of future generations should not be ignored. However, in the absence of a comprehensive dynamic welfare theory, our treatment is bound to be experimental and approximate.

12.2 TECHNICAL DEVELOPMENT AND WELFARE THEORY

Economics has gradually freed itself from the traditional objectivistic concept of welfare which equated welfare with consumption. The widespread practice of equating technical change or development with technical *progress* (because it raises production) must therefore be regarded as a relic of the materialistic interpretation of welfare. This attitude excluded the assessment of the advantages and disadvantages of technical development—a task of paramount importance for both present and future generations—from economic policy;[28] and with R & D expenditure, where any increase is generally welcomed without further examination, no alternative uses for scarce resources were considered.

The attitude that technical development is synonymous with an improvement in welfare has led many economists to put more emphasis on the growth of production than on optimum allocation in a given period. This view and non-economic value judgements about relative merits of growth and resource allocation should be clearly distinguished from a welfare-theory analysis of the economic significance of growth in general and technical change in particular.

Welfare theory is mainly concerned with identifying the prerequisites for an optimum allocation of resources. Pareto's criterion is that allocation is optimal if a change in the allocation cannot achieve an increase in the welfare of one or more individuals without decreasing the welfare of at least one other subject. This criterion of optimal allocation is independent of the distribution of goods, and there is a different Pareto-optimum corresponding to each income distribution.[29]

Although the assessment of a change in the social welfare is thwarted by the impossibility of comparing the utilities of different individuals in cases where a change in allocation increases the welfare of some and decreases the welfare of others, one often speaks of a potential increase in the social welfare if there is a possibility that those who gain in welfare will sufficiently compensate for those who lose. This hypothetical version of the compensation principle is generally used in welfare theory to make a statement, irrespective of the distribution effects, about the influence exerted on social welfare by a change in the allocation of the factors of production.[30]

No systematic examination of the welfare effects of technical development has yet appeared, and it is doubtful whether the predominantly static welfare theory will prove adequate for solving these problems since the nature and rate of technical change affect the welfare not only of present but also of future generations. Using Pareto's criterion, technical development is beneficial if it increases present welfare without reducing future welfare, while it is not beneficial if it reduces present welfare and leaves future welfare unaffected. The use of Pareto's criterion is indeterminate when present and future welfare are affected in opposite directions. The compensation principle is again unhelpful since practically no concrete meaning can be given to the actual or potential compensation of the present generation by the future, or vice versa.[31]

A time element must be introduced into welfare theory if it is to give answers to some questions connected with technical development. Tinbergen, who has made a significant attempt to produce a dynamic welfare theory does not dispense with the foundations laid down by Pareto, but his concept of the dynamic welfare optimum is open to criticism because he introduces value judgments based on personal preferences. His conclusions about the optimal development of welfare are therefore based not so much on the use of an interpersonal comparison of individuals' utilities as on an 'intrapersonal' comparison, i.e. the 'preferences or judgments of the comparer himself'.[32]

For some assessments of the significance of technical development for the present generation, we can borrow Pareto's criterion from static welfare theory. For example, if technical change is neutral in Hicks' sense and therefore leaves income distribution unaffected, the optimum allocation can in principle be determined independently of the distribution. Except for this special case, however, there is generally a relationship between income distribution and the type of technical change. The change in social welfare due to a change in the allocation and distribution of resources brought about by technical development can then be judged in principle by the hypothetical version of the compensation principle, according to whether welfare gains could more than compensate welfare losses. If the application of labour-saving techniques leads to redundancies, the loss of income from work may be compensated by increased company profits. In the case of a capital-saving technique, however, the compensation principle is complicated by the method of writing-off depreciation used and the difficulty of identifying the losers. The application of this principle is therefore sometimes hampered by qualitative problems, and sometimes by the inadequacy of the means to make compensation.

A related use of the compensation principle concerns the effect of a new technique on existing capital assets, introduced into the argument by Pigou.[33] Pigou asked whether a new production method produces a divergence between the individual and the social returns on the investment, since the individual marginal product is not corrected for the accelerated depreciation of the existing capital goods. He concluded that there is no divergence because the disadvantage of the fall in prices, which must be conceded even by those who use obsolete production methods, is fully offset by the advantage of price cuts for consumers. Since the gains could more than compensate the losses the new technique means a potential increase in social welfare, so that the individual and the social returns on investment need not differ from each other. The compensation principle can therefore be applied here without recourse to an interpersonal comparison of utilities.

Pigou then examined whether the introduction of improved production methods could be promoted by safeguarding the new equipment from rapid obsolescence with the appearance of even better machines. He believes it can, but that such a policy has the disadvantage that obsolete production methods then remain in use despite the availability of better ones. Pigou does not say which of these two opposite effects is the greater, but he stresses that it has been assumed that the rate at which new production methods are devised is independent of the rate of their actual application. If this assumption is dropped, and the two rates are considered connected, Pigou believes that the advantages of a protective policy towards technical development do not offset the disadvantages. Although this final conclusion echoes the problems raised by List, the distinction Pigou makes between inventions and applications in a welfare-theory deserves special attention.[34] Machlup endorses this distinction when he deals with the social marginal costs and returns of both research expenditure and innovations,[35] using the growth in national product as a measure of welfare.

Recalling the distinction at the beginning of the book between technical change in a narrow and a broad sense, and examining both increases in technical knowledge and in the applications of that knowledge, we obtain four cases, discussed below in turn. We will not only be concerned with the direct changes in welfare but also with the external effects.

Starting with technical change in the narrow sense (i.e. an increase in technical knowledge), we attribute this change to scientific advance, R & D expenditure and education. The relative importance of these factors has been differently assessed by different authors. Thus, Schmookler[36] believes that there is no close connection between science and the stream of inventions, while Nelson[37] comes to the opposite conclusion. Jewkes[38] has pointed out that science should be more clearly distinguished from technology—for example the early Soviet successes in space exploration resulted in Britain erroneously allocating more money to science rather than technology. But we need not concern ourselves here with a detailed assessment of the relative importance of these various factors. We can assume that all the factors mentioned are involved, and may influence one another. This implies that inventive activity is influenced not only by demand considerations such as the pattern of human wants and the composition of demand— which is Schmookler's view—but also by the supply side, i.e. science and technology. In Rosenberg's words: 'A central problem is to trace out carefully the manner in which differences in the state of development of individual sciences and technologies have influenced the composition of inventive activities.'[39]

The conditions for an optimum allocation relate both to the size and the direction of the generation of technical knowledge. In a highly original work, Arrow deals mostly with the former,[40] and especially with the question how far perfect competition ensures a Pareto-optimal allocation of the means of production for the generation of technical knowledge. He treats technical knowledge as a good, the production of which is particularly uncertain, since it cannot be predicted from the investment in it. The institutions of a free-market economy which take on the risks involved cannot ensure that an optimal allocation is achieved. Arrow believes that, as a result, there is not enough investment in the production of technical knowledge in a perfectly competitive market. The situation improves when a few firms acquire strong monopoly power in the market, since they can then act as their own insurance companies. However, the monopolization of technical knowledge, for example in the form of a patent system, militates against the free availability of technical knowledge which is socially desirable. Arrow's conclusion is that not enough is invested in the production of technical knowledge 'because it is risky, because the product can be appropriated only to a limited extent, and because of increasing returns to use'.[41] One may deduce from this the related conclusion that the government should itself contribute to research.

Demsetz[42] has criticized Arrow's conclusions, saying they are too deeply rooted in a comparison between a theoretically derived ideal state and the actual institutions, whereas—he argues—the task of welfare theory in this field is to compare the allocations in the case of various actual or potential institutions. Demsetz believes that the prerequisites for an optimum should be identified not with reference to some theoretical ideal, but on the basis of the actual characteristics of the prevailing type of production. Whether or not welfare is increased by government investment in the generation of technical knowledge cannot be determined only by the market mechanism, but also requires a comparison between the nature and consequences of government intervention and the results that can be expected without it. If, owing to the preferences expressed in the market, the cost of a risk shifting and sharing system is greater than its advantages, the absence of such a system does not necessarily mean that the allocation is not optimal in this respect, contrary to Arrow's views.

Yamey says that the inventor faces greater risks when dealing with a monopolist, rather than with freely competitive industry, which will certainly reduce his incentive to invent, unless he can make sure that he gets paid afterwards. This problem vanishes if the inventor and the monopolist are both part of the same organization. Using the Arrow-Demsetz model, Kamien and Schwartz have demonstrated that, whatever the internal structure of the industry, the more elastic the market demand curve, the greater the incentive to invent.

Another aspect of Arrow's model has been elucidated by Aislabie,[43] who has pointed out that Arrow defines technical knowledge as a good in order to be able to apply the usual allocation theory to it, although it lacks a number of the characteristics of the other goods that Arrow considers. The cost involved in the diffusion of technical knowledge is relatively small; it is not 'supplied', because it does not have a market price and it is not 'demanded', because little is known about it. Aislabie views technical knowledge as a good by considering the provision of information as a special kind of labour that creates a link between the potential inventor and the potential user of knowledge. This extends the allocation of the means of production to those who deal in technical knowledge.

Aislabie's contribution highlights the complicated character of technical knowledge, which we previously called a potential collective good. It is on balance unsatisfactory to apply traditional allocation theory for individual or collective goods, and the analysis of the implications and causes of the lack of markets for future goods, indicated by Arrow,[44] may prove more fruitful. We shall merely mention here a few aspects that may be of interest for improving the allocation theory.

New products and new methods of production can be divided into quantitative changes, such as the production of the same output at a lower cost, and qualitative changes, such as greater job satisfaction. Schmookler's investigations show that a large demand for certain consumer goods not only influences the allocation of the factors of production in the usual way, but also influences the use of these factors towards the discovery of related consumer goods. It is an important welfare consideration whether this is based on consumers' demands or on autonomous research activity.[45] If the latter is mostly concentrated in firms these do not merely respond to the requirements of consumers, but instead exert a decisive influence on the volume and composition of consumer goods via the nature and extent of their research. The initiative lies with the firms. Consumers may feel some alienation if their 'needs' have been induced by unknown forces.[46]

If production is based on existing consumer preferences, the standard welfare theory can in principle be employed, the results depending on how far technical knowledge is regarded as an individual or collective good. The situation is completely different when the allocation of the means of production is based on the firms' autonomous research activities; consumer preferences are then at least partially influenced by the research work of the firms. It remains an open question whether a wider range of consumer goods in itself means greater welfare.[47] But we can analyse the principal welfare effects by taking into account not only the financial sacrifices (a pseudo-objective factor), but also the subjective welfare effects: thus, the pollution of the environment and the destruction of nature, which may accompany the generation of technical knowledge, affect welfare, although the market mechanism does not reflect this.[48] As regards labour as a factor of production, one may for example ask whether the amount of education aimed at equipping people for research work on new products is justified.[49]

An increase in technical knowledge is a condition for the maintenance or increase of a firm's source of profit and thus its growth, and also of the maintenance of employment. A reservoir of knowledge which the company can tap at a later date is another individual welfare effect. Here an enterprise must decide whether to keep an invention secret or take out a patent on it. From the point of view of society, the former decision may mean the duplication of some research projects and a slower overall expansion of technical knowledge.[50]

Let us now assume that the new technical knowledge concerns techniques, which enable the same amount of a product to be made with fewer inputs, or with the same inputs yielding a higher output. Apart from the welfare effects mentioned in the previous case, the new technique means a potential reduction in the company's costs. It is also important to establish if the new technique requires completely new fixed production equipment and installations, or can be put to use in the form of continuous regular improvements. If the technical change is discontinuous and is therefore embodied in capital goods, the investment decision involved is a far-reaching one. Another important factor is the age of the capital goods that would be replaced. A further consideration is the effect on the welfare of labour—retraining, for example, may involve some financial sacrifice.

If the new technical knowledge is the exclusive property of a firm, the firm's assessment of it may easily differ from, and even conflict with, its general welfare value. The firm will delay the introduction of its new technique if it has just installed new capital goods; if the delay is undesirable to the community as a whole, the government may speed up its introduction with the aid of public funds. This involves complicated problems of assessing conflicting interests, and these problems also have a bearing on the policy on competition. The transfer of knowledge to other firms which are ready to change to new techniques also creates difficult problems for general economic policy (discussed in Section 12.5). Furthermore, employment will be generally affected if labour-saving techniques are introduced over a wide front.

Technical change in the broad sense concerns the change in technical possibilities due to better conditions, education, changes in scale and the discovery of new raw materials and sources of energy. The welfare effects of technical change in the broad sense are connected with changes in the conditions necessary for the application of technical knowledge. Thus, education affects the quality of labour so that methods of production can be put into operation which would otherwise be impracticable. In this sense, the dissemination of technical knowledge—technical change in the broad sense—exerts important welfare effects.

Although the actual increase of technical *knowledge* has been central to the above discussion of welfare effects, it has not been possible to avoid mentioning applications as well. Undesirable external effects, closely connected with the application of new techniques, are not evils which occur independently of human will. The applications of new inventions are decided by men after considering economic and other factors, including— at least in part—the environment.[51]

The introduction of a new technique that rationalizes production quite often causes a certain amount of unemployment, which may be prevented by avoiding large wage increases: in some cases, higher wages and the maintenance of jobs are alternatives, since technical development makes it possible to replace labour when its price becomes relatively high.[52] While the best production method is normally taken to be that which gives the cheapest combination of inputs, in a broader welfare-theoretical treatment the identification of the best production method also includes subjective considerations, such as job satisfaction. If the emphasis is not on the quantitative but on the qualitative changes brought about by the new applications, we can confine the discussion of the welfare effects to labour. An important and typical case is one where a new technique makes unskilled labour redundant while creating vacancies for skilled labour. In such a case, technical development affects also the nature and extent of education. It is important to remember that retraining to a higher level may eventually involve greater job satisfaction.

The welfare effects of technical development and its applications both in the narrow and broad sense basically boil down to the good and bad effects of continuous economic growth. The effect on income distribution is a special problem here, for both wages and profits are considerably affected by the various ways new technical knowledge can be applied.

12.3 MODELS OF OPTIMUM GROWTH

The previous section dealt with the effect of technical development and its applications on the life of the present generation. We now examine dynamic models, in which optimum development is viewed over time. The significance of this is admittedly limited, since technical development has not yet affected the future welfare functions involved, but the different views that have been put forward are still important for understanding allocation in the course of time.

The distinction Hennipman makes in the article quoted at the beginning of this chapter between a positive and normative interpretation of the interpersonal comparison of utility is important for an understanding of these optimum growth models. The models cannot be based on a positive interpersonal comparison, since such a comparision cannot be derived from an economic analysis. The models must instead be ultimately based on ideological and political value judgments, often expressed as interpersonal and intertemporal comparisons of utilities.

The theory of optimum economic growth—known until recently as the theory of the optimum rate of saving[53]— generally starts from some assumptions about production methods and the growth of population, and assumes that the utility derived from consumption over the course of time is as great as possible, i.e. that $\sum_1^T u(c_t, t)$ is maximized. Finding the best allocation of production between consumer and capital goods over the course of time thus involves solving a maximization problem with certain boundary conditions.

An intertemporal utility function is a special dynamic version of Bergson's[54] welfare function. The word 'welfare' may give the impression that welfare is used in the general and formal sense in these growth models. But this is not the case. Each model entails one or more special formulations of the objective function, giving rise to as many welfare assessments, each with a special, though only a partial, significance. Thus, Tinbergen and Bos's[55] statement that the maximization of utility should be rated higher than that of consumption, which in turn should be rated higher than that of incomes, is based on a 'broad social judgment'.[56] Even when it is only utility maximization that is considered—as is customary nowadays[57]—it is difficult to claim that welfare is fully accounted for by attributing utility exclusively to the stream of consumer goods—it depends, amongst other things, on the volume and composition of the stock of capital goods.[58] The 'time horizon' in optimum-growth models presents a special difficulty in this connection: with a finite time horizon it is difficult to assess the stock of capital goods at the end of the period, while with an infinite time horizon ($T \rightarrow \infty$) the existence of an optimum solution is very doubtful, because the utilities do not yield a convergent series when added over time.[59] In either case different evaluations are implied. The way the aggregation of the utility functions in time is interpreted indicates that it is an intrapersonal rather than an interpersonal comparison of utilities that is involved, because the aggregate utility function can reflect only the present generation's preferences about present and future consumption, and it is impossible to compare these with the utility a future generation is likely to derive from consumption. The aggregate utility function reflects not only the choice of the arguments to be included in the objective function, but also the method of assessing the utility effect on the various generations.[60] Asimakopoulos[61] also distinguishes between the intertemporal allocation of the stream of consumer goods and the distribution of the goods over various age groups within the same period. The social welfare functions of these two situations need not be the same. The intertemporal dependence of the consumer preferences has been examined more closely by Samuelson[62] and by Ryder and Heal.[63] Apart from these welfare-theoretical complications, there are also particular mathematical problems concerning the criteria for selecting the best programme[64] and establishing the dynamics of optimal equilibrium growth.65 Peleg and Ryder[66] have recently solved the problem for a multi-sector model, in terms of axiomatic production theory.

Bearing in mind the limited welfare significance of the theory of optimum growth, these models do form a useful contribution to economics. It is doubtful if a growth policy[67] that starts from the subjective factors can ever be based on this theory; but by examining the consequences of various assumptions made about the state of technology, capital formation and the growth of population on the one hand, and the choice of the criteria for maximum welfare on the other, some important, though incomplete, insight is gained which can help with the planning of present and future production—the central problem in the theory of optimum growth.[68] An interesting illustration of this can be seen in Forster's attempt to incorporate the pollution problem in an optimal-growth model.[69] We now examine the way technical development is incorporated into such models.

The theoretical models began with the incorporation of the standard case of exogenous technical change in the narrow sense, in which change is expressed by a shift in the production function over time. One of the first examples of this is Shell's treatment,[70] which starts with a Hicks-neutral technical change with a fixed non-negative coefficient ϱ (≥ 0), a well-behaved neo-classical production function, and certain assumptions about population growth and capital formation. Shell then develops a model in which the rate of saving, $s(t)$, over time is determined in such a way that per capita consumption is maximized. In this form, the problem is closely related to the

'Golden Rule of Accumulation',[71] the original formulation of which states that, under the same conditions and in the absence of technical change ($\varrho = 0$), per capita consumption is at a maximum when a growth path is chosen where the marginal physical product of capital is equal to the growth rate. Shell derives a modification of this for the case $\varrho > 0$. In terms of the variants of the theory of optimum growth that are based on Ramsey's ideas, we are dealing here with a special case, since the direct maximization of the per capita consumption C/L amounts to the introduction of a linear utility function. If general utility functions are taken as a basis, then the growth path for which the Golden Rule holds need not be optimum in the sense of maximizing the utility integral:[72]

$$\int_0^T U(C/L).dt$$

Mirrlees has used such an objective function, though he mainly restricts his analysis to the case of a utility function with a constant elasticity, and he assumes, besides, neutral technical change in Harrod's sense. The contribution of Shell and Mirrlees can be summed up by saying that these authors have determined the optimum rate of saving where, given the productive possibilities, various types of technical change are considered, and as such these developments of the theory of optimum growth follow the tradition started by Ramsey. The same applies to Goldar's extension of Shell's study to the case of embodied technical change.[73]

Models which emphasize the endogenous character of technical change have also been combined with optimum growth theory. Sheshinski[74] has developed a model that includes the accumulation of experience in Arrow's sense amongst the technical constraints of optimum growth. Sheshinski studies both the case of a linear utility function and a non-linear one. Phelps has fitted Sheshinski's analysis of the determination of the optimum rate of technical change into a model in which he arrives at a Golden Rule on the basis of the maximization of the per capita consumption.[75] Since the studies by Sheshinski and Phelps also deal with the determination of the optimum rate of saving, they too represent variants of Ramsey's traditional model, even though Phelps maximizes the rate of technical change by choosing from the available labour-augmenting and capital-augmenting types. However, there is a danger here of a confusion of terminology: we first encounter the optimum rate of saving, then the optimum rate of growth, and now a tendency to speak of optimum technical change. Thus, Nordhaus calls a model with the same structure as Ramsey's a theory for establishing the optimum direction of technical change. The rate of saving $s(t)$ and the direction of technical change, represented by an innovation curve, are determined simultaneously to maximize per capita consumption; in other words, a linear utility function is again maximized, with the constraints being provided by the choice of capital formation and the nature and rate of technical change. The question is therefore at the same time one of the optimum rate of saving and the technical change.

The danger in using the terms 'optimum rate of saving', 'optimum growth' and 'optimum technical change' is that they suggest that the situation is optimal from a general welfare point of view.[76] But the incorporation of exogenous and endogenous technical change into the theory of optimum economic growth (in which the nature and rate of technical change are so chosen that the satisfaction of needs by the stream of consumer goods is maximized) does not mean that the technical change can be regarded as optimal from the welfare point of view, which requires non-economic value judgments. Another objection to the term 'optimum technical change' is that technical change in these discussions is not an end but a means, whereas in the models of optimum economic growth, growth and technical change are actually both ends and means.

12.4 OPTIMUM TECHNICAL CHANGE

To examine more closely the meaning of optimum technical change in the economic literature, let us look at the connection between optimum intertemporal allocation and the choice of technical possibilities; to do this we return to the wage-interest curves belonging to different techniques in a simple macroeconomic model of equilibrium growth. Besides the wage-interest curve $w = f(r)$, we take a corresponding ('dual') relationship, between the per capita consumption c and the rate of growth of capital g, represented by the function $c = F(g)$. This duality implies that if the rate of interest is equal to the rate of growth of capital the real wage is equal to per capita consumption.

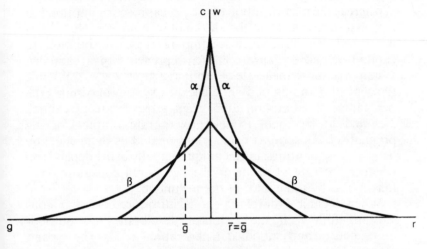

Figure 8

A direct consequence of this duality concerns the Golden Rule mentioned in the previous section. Figure 8 shows the wage-interest curve and the corresponding c-g curve for two techniques α and β. With a rate of growth $g = \bar{g}$, per capita consumption c is maximized by choosing technique α. This technique will be chosen via the price mechanism in perfect competition if the interest r is equal to the growth rate g. This representation of the Golden Rule theorem illustrates the connection between the optimum growth models mentioned above and the choice of technique. The maximization of the stream of consumer goods in time by choosing a growth path g such that $g = \bar{r}$ is reduced to the choice of a technique where the real wage is maximized if $\bar{r} = g$. However, this reveals at the same time that the theorem is more formal than substantial from an economic point of view, since it concerns mainly the technical characteristics of the economic process, and there is only an apparent choice here based on consumers' preferences.[77]

This can be further developed by the assumption that the rate of interest r is not equal to the growth rate g. Irrespective of how r and g are determined, the choice of technique then depends entirely on the objectives discerned in the description of the technical possibilities. At a relatively high value of r, the maximization of the real wage w leads to the choice of technique β, whereas the maximization of the per capita consumption c at a growth rate g leads to the choice of technique α. Nuti[78] has developed a model in which he compares these two situations to capitalism and socialism. The entrepreneurs in a capitalist system maximize their rate of profit at a given wage level, while the central planning body in a socialist system maximizes the per capita consumption at a given growth rate of capital, assumed to be equal to the growth rate of the working population. As soon as r and g are treated as independently determined quantities, the technique actually in use depends on whether the rate of profit or the per capita consumption is maximized by the choice of the technique.

When technical development is introduced, similar conclusions follow, if the technical development is autonomous, neutral and embodied in new capital goods. The w/r and c/g curves then exhibit an identical shift, while the character of equilibrium growth is retained in the model. Optimum technical development is again limited by the two objectives derived from the description of technical development using wage-interest and consumption-growth curves.

From a welfare-theoretical point of view, this analysis is of limited significance since it has so far not allowed for other objectives than those mentioned above. The functions $c = F(g)$ and $w = f(r)$ as such describe only the technical possibilities, so that the choice of the best technique overall generally needs the introduction of a social welfare function.

The two objectives discussed above can of course be obtained from this welfare function, but there are other welfare functions besides the intertemporal utility functions discussed, incorporating other factors that influence welfare. If there is a large number of technical possibilities, several optima can be proposed, depending on the relationship between the welfare function and these possibilities. Techniques that cause the least environmental pollution will often be different from those which are the most profitable.

Such an analysis could start with the construction of a social welfare function from the relationship between per capita consumption and the growth rate. The counterpart of the maximization of the real wage rate, given the rate of profit, being the maximization of per capita consumption, given the rate of growth of the system. It is then a small step to conceive of social welfare as a function of c and another step to broaden the welfare function to include other factors. The traditional way of deriving the envelope is based on the assumption that the social welfare function is $U(c) = c$. The same result can be derived if we use the more general social welfare function $U = U(c)$. A still more general social welfare function considers utility as a function of the wage share α and per capita consumption c. The significance of this maximization of social welfare for the set of optimal techniques can then be studied. If we apply the social welfare function $U(c, \alpha)$, not only is the functional income distribution determined in an optimal way by choosing a certain technique, but also per capita consumption. The interesting result is that if we assume that the rate of growth g is given exogenously, the optimal technique will again depend on the rate of profit r.

The quasi-envelope may reveal a generalization of the reswitching phenomena in so far as techniques which are chosen at certain levels of the rate of profit and the rate of growth to maximize social welfare are again the best choice at other levels after being replaced at intermediate levels of r and g.

We therefore come to what is perhaps the most interesting feature of this analysis, the fact that—contrary to Nuti's analysis—it depends on the specification of the social welfare function whether or not the consumption per head is determined by g, or by g and r. It is now no longer possible to associate capitalism with the (w, r)-domain and socialism with the (c, g)-domain. Even in a capitalist society the criterion of an equitable distribution of income may play a fundamental role in the concept of social welfare. As soon as this is the case, the discussion cannot be framed in terms of fixing g or r and consequently choosing an optimal technique, but, rather, the outcome is a matter of knowing g *and* r.[79]

Considerable complications arise if technical development in all its variety is introduced, instead of a certain set of technical possibilities or their uniform expansion. If technical change is embodied in new capital goods, there will be a difference between the cost of the old and the new capital goods, not reflected in the wage-interest curves. As new consumer goods appear on the market due to technical development, the duality of the functions $c = F(g)$ and $w = f(r)$ disappears, because it is based on a given composite of consumer goods. Furthermore, when the technical change is no longer assumed to be neutral the concept of equilibrium disappears. As a result of the endogenous character of technical development, the choice of the direction and rate of the latter itself becomes subject to welfare considerations. Even if we ignore the re-adoption of previous techniques, we must conclude—in view of the heterogeneous nature of technical development—that an identification of welfare-optimal technical development cannot be based on a comparison of the wage-interest curves and the corresponding consumption-growth curves.

On the one hand, the objectives derived from the construction of these curves do not follow from an exogenously fixed welfare function, and on the other hand the introduction of such welfare functions implies the recognition of a variety of factors that influence welfare. This is particularly true if a monopolistic market is assumed instead of perfect competition, for the different behaviour patterns of the entrepreneurs must then be also taken into account.

However we can clarify the concept of optimum technical development by a closer examination of the influence of economic policy.

12.5 ECONOMIC POLICY
If we define economic policy in terms of ends and means, equilibrium technical development should be added to the aims of economic policy. This makes the rate of technical development part of government policy, so that technical development is no longer regarded as an autonomous phenomenon due to God and engineers, but one that is in principle amenable to regulation and control.

This subordinates technical development to subjective values and preferences. It is a semantic rather than a conceptual question whether the type and rate of technical development are regarded as aims or as means of the government; in some cases, a particular technical devélopment can be viewed as the aim of the government or as a means for achieving further aims. The important point is that technical development should be considered as a process that can at least be influenced by the government and has profound consequences for welfare.

The realization that, from the social point of view, the market mechanism often provides imperfect guidance for optimum technical development emphasizes the need for a government policy worked out by the appropriate decision-making processes. The state is then not only the guardian of the common good, but also the organizer of decision-making in society in the broadest sense. It is the special duty of the state to look after the interests of future generations, and it may prohibit by law the application of new technical inventions if they are deemed undesirable from this point of view. This is particularly important where these applications are irreversible.

The solution of environmental problems calls for the organization of close consultation between the government, the business sector and consumers. The government by itself cannot preserve the environment, because it lacks the necessary technical knowledge. Companies can no longer avoid responsibility for the often serious external disadvantages of their production methods, but it is not possible for them to assess these wider disadvantages against advantages such as the provision of employment and the contribution to the balance of payments; the market position of most companies also limits the extent to which they themselves can bear the cost of environmentally desirable production methods—although the government should encourage management to be aware of the wider needs of society. Consumers have an influence—via the market—on the supply of goods, but the individual consumer cannot assess all the factors behind the production of these goods. Since the public and private sector have become increasingly interdependent, a consultative body should be established and provided with the necessary technical knowledge together with information about the welfare effects of the various solutions, so that it can furnish a basis for decisions which take into account both present and future generations.

The coordinating and organizing function of the government can be extended to the whole field of technical development. No general answers independent of the particular circumstances can be given to questions such as whether the government should restrict or promote the monopolization of technical knowledge by companies, whether it should encourage the transfer of technical knowledge from one company to the other, whether it should curb the growth of a company because of external disadvantages despite favourable economies of scale, or whether the geographical and sectoral diffusion of technical development is desirable. The amount it spends on the development of technical knowledge alone (for example on education) justifies a more integrated management of the public and private sector as regards technical development.[80] The more that a company's technical invention is the fruit of direct or indirect public expenditure, the less defensible is its secrecy.[81]

A uniform economic policy does not fit the heterogeneous character of technical development and its consequences. Decisions about the extension of technical knowledge and about its diffusion and application have far-reaching consequences for the number and type of jobs available, the environment, real income per head and the welfare of future generations. The assessment of all these effects requires a microeconomic approach to do justice to the variety of welfare effects which differ with time and place.

My personal view is that both the extension of technical knowledge and the application of inventions should be part of economic policy. Modifications to technical knowledge by changing the extent and trend of research work should lead to a more rapid diffusion of the results. We are not only dealing with technical knowledge that yields an immediate financial return;[82] we must also bear in mind the effects of technical change which appear slowly or indirectly. A form of organization must be chosen that permits consultation between the public and the private sector. Very large companies with considerable monopoly power may require the international coordination of policy. Examination of these companies would reveal to the public the extent to which monopolistic profit is necessary for research. It might also reveal that some companies contribute too little to the generation of technical knowledge, because their profits are insufficient or wrongly allocated. Smaller firms should also be taken into account to avoid lopsided industrial development, itself partly due to technical change.[83]

The applications of technical development provide a more concrete object for economic policy than the increase of technical knowledge. The introduction of new methods and new products can be postponed for various lengths of time, or geographical strings attached; while the effect of new techniques on the labour market and the environment is clearly important. The government should be prepared to promote innovation in companies and industries where the slow development of technology has unfavourable welfare effects. Environmentally clean techniques can be encouraged by taxes and fines or the introduction of prohibiting laws. Education in general, and retraining in particular, can be used as instruments of a labour policy that takes technical development into account.

In a simpler world, economic decisions were easy. Everyone abided by simple accounting procedures which revealed maximum profits and minimum overheads. It was a private matter for private firms, and no-one, except a few radicals, questioned it. Decisions about technological innovations or the modification of existing technology were based only on financial considerations, and it was the managers alone who were qualified and responsible for making such decisions.

In our more complex modern world this is no longer the case. Managerial decisions must now also take into account the welfare of the general public, the consequences for the whole economy, for the labour market, the availability and scarcity of raw materials including energy, and the effect on the environment. A great deal of knowledge has been accumulated relating to all these, and other, aspects of technology, and it is imperative that such knowledge be utilized in making economic decisions.

For example, let us assume that a large firm is faced with a

number of alternative techniques for improving its operation—
and in the modern world, there are always alternatives.
Financially, it may be clear which alternative produces the most
profits. But other considerations show us that technique A may
be more labour-saving than technique B; and technique C may
require a different type of labour. Technique D may be
preferred because of its favourable effects on working
conditions, while technique E is the choice of the environ-
mentalists. Techniques do not only differ from one another in
these respects, they also differ with respect to their effects on
energy, raw materials, health, and so on.

It is obvious that there are public welfare effects resulting
from the ostensibly 'private' economic decision of the company
faced with the above choice. Moreover, the effects are
interwoven and have many aspects which are not readily·
apparent. For instance, the noise level of a machine in a factory
mainly affects the individual worker who is subjected to it.
Pollution has to do with public welfare effects of the choice of
techniques. As for the effect on employment, the private and
public aspects are connected. Thus, public welfare cannot be
ignored, and governments are increasingly concerned to ensure
that welfare does not suffer by technological revolution.

From the different aspects of the choice of techniques
described above, it follows that the choice cannot be optimal
and efficient if it is based only on private welfare effects of a
primarily financial character and is made by one or more
managers of the firm. Society has therefore to organize a
decision-making process which enables us to take into account
public welfare effects and then private welfare effects, which
cannot be derived exclusively from the market and the price
mechanism.

Within the firm this implies a democratization of the
decision-making process in such a way that all concerned can
express their feelings about the different welfare effects,
research and technology. It also implies that the task of the
manager is not only to make decisions, but also to *organize*
them. The welfare effects, once ascertained, should be assessed
by the individual company or economic enterprise and
implemented or rejected on the basis of government policy. A
policy can thus be shaped in such a way that people control

technology rather than vice versa. Failure to do this will involve other consequences besides economic ones.

All this adds up to the fact that decisions about the nature and rate of technical development should not be exclusively in the hands of companies, in view of their far-reaching consequences for society as a whole.[84] In Schumacher's words, we need technical development with a human face.[85]

12.6 CONCLUSIONS

The central theme of this chapter has been the belief that technical development needs guidance and control. Technical development upsets equilibria (if these are attained at all), and aggravates disequilibrium situations, mainly through the economics of scale it produces. In this respect, it is a source of continuous uncertainty. It threatens to make man a puppet in the hands of unknown forces. If we are to maintain and strengthen our legal and social system, based, as it is, on a free society, we must devise a way of controlling this process, for it may otherwise overpower us.

This control can be more firmly based on a microeconomic than on a macroeconomic analysis of technical development. The macroeconomic optimum-growth models discussed show technical change as an uncomplicated and trouble-free phenomenon—a picture that does not fully correspond to reality, for it ignores effects of technical development, such as the changes in the market power of companies, the organizational problems within the companies, and the changes in the employment situation. Therefore a microeconomic approach is mainly required for analysing the welfare effects of technical development. This links up with the proposal that the guidance and control of technical development should be carried out in close consultation with individual companies.

The active guidance and control of technical development by the central government is fully justified by the effects of technical change on employment and the environment. The need for participation can be satisfied by designing suitable procedures for decision-making. Since decision-making without information and understanding is pointless, detailed policy models must be constructed in which various alternatives are given, as far as possible, in precise and quantitative form. The development of cost-benefit analysis[86] is a step in the right direction. It is possible that we may thus achieve the eventual convergence of initially different value judgments.[87]

The fact that pollution and the depletion of energy resources affect the whole world raises difficult problems in connection with technical development. The first report of the Club of Rome,[88] which created a sensation, is open to considerable criticism.[89] In any period the assumption of a permanent exponential growth can lead to the conclusion that catastrophe is unavoidable.[90] No econometric verification was offered in that report, nor was technical development given its proper due. Mishan has said that the pioneering work done by Jay Forrester and Meadows was only a crude beginning.[91] The lack of feedback mechanisms in the overall dynamic system of the Club of Rome has stimulated intense theoretical and empirical examination on the corrective potential of prices and of technical development. The negative external effects of economic growth on the environment are partly reflected in prices in so far as people are prepared to sacrifice some resources to remedy these negative effects, and in so far as the requirement of an unpolluted and less noisy environment is translated into prices.[92] To the extent that these automatic mechanisms turn out to be inadequate, some form of public policy is needed, based on a comparison of the advantages and disadvantages of economic growth for society as a whole. In the case of finite resources, the price mechanism can play a role in checking the unlimited and thoughtless use of natural resources. Indeed, as Solow has suggested, the market for exhaustible resources 'might be one of the places in the economy where some sort of organized indicative planning could play a constructive role'.[93]

Paradoxically, it is the pollution of air[94] and water, often caused by technology, that is most likely to be remedied by new techniques.[95] While many economists have noted that the Club of Rome ignored technical change, they do not agree on the effect of the latter on the so-called Doomsday Model. Thus, Mishan[96] doubts that technical development can solve the problems of unlimited exponential growth, while Kranzberg[97] firmly believes that technical development 'can provide for the future' if it is guided by public control. My own view inclines to the latter, optimistic outlook, but one must acknowledge the useful social impact of those who—like Jevons in the last century[98]—hold more pessimistic views. A world-wide policy on environment and energy is obviously needed, but one must be cautious about drawing over-pessimistic conclusions.

In physics, the Universe is regarded as a closed system in which irreversible processes are accompanied by an increase in entropy, with low-entropy forms of energy such as the force of gravity being converted into higher-entropy forms, such as heat; this process finally leading to completely degraded forms of energy and the annihilation of the Universe. The stream of energy from the sun is part of this process. While we cannot escape the long-term thermo-dynamic conclusion that life on earth will come to an end, the time-scale involved exceeds human imagination, so that these considerations cannot be the basis of realistic pessimism.

Owing to the introduction of energy from the sun, the earth's biosphere forms an open system, where order can arise from disorder and a large number of cyclic processes can be maintained. This has led more or less randomly to the accumulation of supplies of chemical energy, but owing to the undirected nature of technical development these energy sources are now being depleted relatively quickly.[99] We are living off a finite stock of energy. Much energy is wasted in its own transport, use and conversion from one form to another, and production has become dependent on resources formed millions of years ago.

In so far as nuclear energy is based on the splitting of uranium and plutonium, it also represents a finite source of energy. But this form of energy is an important temporary substitute for fossil fuels such as coal and oil, supplies of which are rapidly running out, particularly if use is made of breeder reactors, which produce new fuel while producing energy. However, the use of this energy source raises difficult problems of thermal pollution and the disposal of radioactive waste. Atomic energy can also be obtained from nuclear fusion in which light nuclei are combined. This energy source is in principle inexhaustible, but the technology required for harnessing it is so complex that it will be some time before it contributes to the solution of the energy problem.

Nuclear fusion brings us to processes involved in the production of solar energy. It can be argued that some problems so strongly emphasized by the Club of Rome should be solved by the direct utilization of the radiation from the sun, together with other natural forces, such as wind power, water power and photosynthesis, which do not introduce any heat into the atmosphere. The use of these forms of energy means that the conversion of various forms of energy into heat would be slowed down—which would make it possible to use the energy sensibly and profitably, and which fits in with the natural cycles. As a result we would be living on interest rather than capital.

Technical development should primarily be geared to these invariant natural energy systems, and attempts should be made to identify the most suitable type of energy conversion, find more efficient ways of using energy, and solve the problems of the transport, storage and location of heat released in bulk, together with the problem of thermal pollution. In the meantime it should be borne in mind that the most important sources of current energy, namely the irreplaceable fossil fuels, will run out, so that there is a theoretical upper limit to production. The gravity of the present situation mainly depends on how close we have already come to this limit.[100]

It can be assumed that circumstances will force mankind to use solar energy and other invariant energy systems as sources of power for industrial and mechanical processes. However, technical constraints may make it necessary to develop nuclear energy—both fission and fusion—as an intermediate stage, though this has environmental consequences. Although these considerations indicate that it should be possible to ensure the long-term energy supply of mankind, growth will still be limited by other factors, such as the finite size of the earth, the resources of raw materials and ecological principles.

13 The shifting frontiers

For if the changing character of man's wants tends to govern the growth of knowledge, then all men must accept some measure of responsibility for what happens next.

Schmookler[1]

Veblen's picture[2] of the effect of mechanization on economic, social and cultural life at the beginning of this century illustrates how horizons change with technical development. For him the machine was not an incidental product of economic affairs but a starting point, and an expression of, the life of the entire society. Mechanization in fact exacerbates scarcity rather than eliminates it. Technical development creates new needs by satisfying old ones. The stream of consumer goods is constantly renewed. The introduction of new techniques requires more and more preparation, guidance and control, particularly over its social consequences.

The justification of the broad definition of technical change—the creation of new technical possibilities—lies to a great extent in the fact that it does not interrupt the investigation of the causes and effects of technical development at an arbitrary point. Technical development and the state of technology have long been included among the non-economic assumptions of economic science, on the grounds that the creation of new technical possibilities is so remote from economic factors that the usual tools of economics are unsuitable for analysing it. As regards the creative and inventive element involved, this opinion has a large grain of truth; economic science is not suitable for explaining the unforeseen.

The way in which scarce resources are used depends on the technical possibilities, but at the same time it influences the creation of new techniques. Qualitative and quantitative differences in the use of scarce resources are important for the nature and extent of technical change, although they certainly do not completely account for the latter. Since technical development depends on the use of scarce resources, economics as a whole can no longer regard technical development and the state of technology as given, non-economic quantities.

The old frontiers of economics are changing, but where will the new ones lie? The answer is not easy, because of the effect of technical development on other non-economic factors, such as law and social structure. If consumer preferences could be considered independent of technology, it would be logical to regard them as external data for economics, and they could explain technical change. However, this produces an oversimplified picture in two respects.

In the first place, new products alter the pattern of demand. Irrespective of the extent to which consumer preferences affect the nature and rate of technical development, consumer preferences are certainly changed by the appearance of new products. The prevailing view that technical development affects particularly the supply side of the economic process is one-sided, for in reality technical development and consumer's requirements influence each other.

In the second place, the use of scarce resources is influenced by individuals in several other roles than as a consumer, namely as owners of the factors of production, as producers, and as participants in collective decision-making. On these grounds, it is more logical to regard only subjective preferences about the use of scarce resources as given factors for economic theory. The analysis of technical development then comes within the scope of economics, even though economics cannot be expected to give a full description of such a multifaceted and fundamental process. Technical development is therefore an allocation-dependent process, based ultimately on the conscious and subconscious wishes and preferences of human beings.

However, insofar as these preferences in turn depend on technical development, this approach does not reveal a straightforward cause-and-effect type of technical development. Such an explanation is also at odds with the fact that the supply side plays a role in the level and structure of inventive activity. Technical development presents not a closed field of force, in the mechanical sense, but an open-ended succession, in the biological sense.

While it is an acceptable provisional method to derive static and dynamic equilibria as an expression of the given quality of the endogenous economic variables, when technical development is introduced equilibrium is replaced by constant disequilibria. Market equilibria can be derived at a given state of technology because both the demand and the supply side of the system are then closed. But if the state of technology is not something that is exogenously determined, a factor that offers an explanation becomes a factor that requires one, on both sides of the market. The open-ended system of interaction means that an explanation of the state of technology at a given set of prices is as much needed as the explanation of prices at a given state of technology. It seems logical that in a longer perspective of technical development the concept of a general price equilibrium will eventually be replaced by one of continuously expanding technical possibilities, provisionally as something that depends on economic factors, and later as something that has a reciprocal relationship with these factors. Gintis follows this line of thought when he attacks the assumption that preferences are given data, independent of social development.[3]

The momentum of technical development is increasing and has already affected attitudes in society profoundly. Religious attitudes have changed radically; ever-changing demands are placed on education; international political relationships are changing with the change in military technology; the pressure on people involved in production increases; managers cannot keep up with the ever-increasing rate of change; the need for incessant innovations is intensified (although the allocation of research funds between the development of products and of production methods becomes more and more difficult). Technical development, as it were, rampages through society, looking for larger and larger footholds; which leads to the 'appearance of an economy of collective action and large economic power groups'.[4] Technical development goes from strength to strength and knows no inherent limits. It encompasses both good and evil.

The view that technical development is ultimately the outcome of human activities, preferences and decisions implies that economic and social policy can influence it. Now that technical development can be viewed not as a datum external to economics but as an endogenous element of economic theory and policy, the frontiers of economics have definitely shifted.

But frontiers can only shift if they exist. Although history provides an ever-changing picture of technical development, with an anything but equilibrium pattern and a course of events characterized by 'technologically induced traverses, disequilibrium transitions between successive growth paths',[5] yet there is no plurality without unity, no interruption without continuity. Technical development is not so heterogeneous and chaotic that its present bears no relation to its past. The dynamics of economic life consists in the fact that to a certain extent the same pictures recur.

Such paradoxes are not alien to the field of technical change. When rapid technical change is expected, this expectation itself will slow down the application of new techniques, because in a rapidly changing world people tend to adopt a wait-and-see policy, which might in fact be desirable both from the private and from the public point of view.[6]

In the next few decades, economics will return to its traditional role of explaining the qualitative and quantitative increase in the satisfaction of human needs and the study of the best ways of handling scarce resources. This will bring to an end the Keynesian era. The macroeconomic preoccupation with expenditure will give way to the microeconomic analysis of technical innovation. Keynes always avoided a systematic study of technical development, but he did say[7] that its absence causes secular stagnation. There seems no danger of this. The pool of technological knowledge is now so enormous. Kahn and Wiener even expect that cybernetics and automation will be so advanced by the end of the century that, for example, routine household chores will be done by robots.[8] However, we cannot conclude from this that—as Keynes said in an optimistic moment[9]—'the economic problem may be solved' around the year 2030. It is more likely that the human mind will be faced with greater and greater problems of reconciling increasing needs and limited resources.

New methods of forecasting technical change are being developed, and thought is being devoted to the economic effects of anticipating such change.[10] Such efforts gain in importance as the process of technological change—operating through the medium of multinational companies—is less and less constrained by national frontiers. Hence, establishing criteria for economic policy involving technological change can no longer be done on a national basis. Individual Western European countries must recognize that these economic processes are supra-national in nature, and call for a concerted European policy.

It must be recognized that change involving technological development extends beyond the frontiers of economics. It embraces all aspects of human activity. With such a recognition, and with broadly based social and scientific research, policies will be based on more than opinions, and crude dogmatism will be tempered by knowledge. Only in this way can technical development be managed and controlled so that it will be the handmaiden of human progress and not its master.

Notes

The following abbreviations are used in the Notes:

AER	*American Economic Review*
CJEPS	*Canadian Journal of Economics and Political Science*
EJ	*Economic Journal*
HPE	*History of Political Economy*
IER	*International Economic Review*
JEL	*Journal of Economic Literature*
JET	*Journal of Economic Theory*
JPE	*Journal of Political Economy*
OEP	*Oxford Economic Papers*
QJE	*Quarterly Journal of Economics*
RE and S	*Review of Economics and Statistics*
RES	*Review of Economic Studies*
SEJ	*Southern Economic Journal*

CHAPTER 1: INTRODUCTION

1 D. Dewey, *Modern Capital Theory* (New York, 1965), p. 140.

CHAPTER 2: THE CLASSICAL ECONOMISTS

1 R. Whately, *Introductory Lectures on Political Economy* (London, 1831), p. 173.

2 J. Steuart, *An Inquiry into the Principles of Political Economy: Being an Essay on the Science of Domestic Policy in Free Nations*, 1st edn (London, 1767).

3 A. Smith, *An Inquiry into the Nature and Causes of the Wealth of Nations* 1st edn (London, 1776), vols 1 and 2. The new edition (London, 1976), ed. R. H. Campbell, A. S. Skinner and W. B. Todd, is recommended.

4 In this connection some general information is to be found in S. R. Sen, *The Economics of Sir James Steuart* (London, 1957), pp. 130–54.

5 Steuart, op. cit., p. 295.

6 Freeman has recently pointed out that, according to Smith, the social advantages of the division of labour are less important if insufficient attention is paid to education. R. D. Freeman, 'Adam Smith, Education and Laissez-Faire', *HPE*, 1969, 173–86. See also R. L. Meek and A. S. Skinner, 'The Development of Adam Smith's Ideas on the Division of Labour', *EJ*, 1973, 1094–1116.

7 Smith, op. cit., Book I, Chapter I, pp. 20–1 (in the 1st edn, pp. 5–12). Some recently discovered lecture notes of Smith's students have thrown

new light on Smith's ideas about the division of labour. See R. L. Meek and A. S. Skinner in note 6 above. See also N. Rosenberg, 'Adam Smith on the Division of Labour: Two Views or One', *Economica*, 1965, 127–39.

8 Ibid., p. 298. This well-known quotation comes from the Chapter 'Conclusion of the Mercantile System', first appearing in the third edition (1784).

9 S. Hollander, *The Economics of Adam Smith* (London, 1973), pp. 208–41.

10 G. B. Richardson, 'Adam Smith on Competition and Increasing Returns' in A. S. Skinner and T. Wilson (eds), *Essays on Adam Smith* (Oxford, 1975), pp. 350–60.

11 It is curious that Smith never mentions Steuart's work, although at least the first volume of his two-volume work was found amongst his books (J. Bonar, *A Catalogue of the Library of Adam Smith* (London, 1884), p. 109; cf. also the somewhat inaccurate description of H. Mizuta, *Adam Smith's Library* (Cambridge, 1967), p. 143. McCulloch's assertion that Smith was well acquainted with Steuart's *Principles* is therefore presumably not entirely correct. Cf. McCulloch's introduction to his edition of *The Wealth of Nations* (Edinburgh, 1889), p. xxxvii, and J. R. McCulloch, *The Literature of Political Economy, a Classified Catalogue* (London, 1845), p. 11.

12 See, however, H. Barkai, 'A Formal Outline of a Smithian Growth Model', *QJE*, 1969, 396–414. This author has constructed a model from Smith's theory, about the important role of technology in Smith's work.

13 *Essay on the Application of Capital to Land, with Observations shewing the Impolicy of Any Great Restriction of the Importation of Corn, and that the Bounty of 1688 did not Lower the Price of it,* By a Fellow of University College, Oxford, London, 1815. See also W. D. Grampp, 'Edward West Reconsidered', *HPE*, 1970, 316–43.

14 Cf. also G. J. Stigler, *Essays in the History of Economics* (Chicago, 1965), pp. 175 and 176. Other important works are E. Cannan, *A History of the Theories of Production and Distribution in English Political Economy from 1776 to 1848* (London, 1893), pp. 157–60, and J. R. Commons, *The Distribution of Wealth* (New York, 1893), particularly pp. 116–239.

15 T. R. Malthus, *An Inquiry into the Nature and Progress of Rent and the Principles by which it is Regulated* (London, 1815). D. Ricardo, *An Essay on the Influence of a Low Price of Corn on the Profits of Stock* (London, 1815). On Ricardo's publication, see S. Hollander, 'Ricardo's Analysis of the Profit Rate, 1813–1815', *Economica*, 1973, 260–82; J. Eatwell, 'The Interpretations of Ricardo's Essay on Profits', *Economica*, 1975, 182–7; and S. Hollander, 'Ricardo and the Corn Profit Model: Reply to Eatwell', *Economica*, 1975, 188–202.

16 J. Bonar, *Malthus and his Work* (London, 1885), p. 235.

17 Malthus, op. cit., p. 38. M. Blaug is wrong in saying that 'all classical writers' have labour-saving new machines in mind. See M. Blaug, *Ricardian Economics* (New Haven, 1958), p. 66. Malthus expressly speaks, both in this pamphlet and in his *Principles* (from 1820), of '. . . a

machine . . . which will produce more finished work with less labour and capital than before' (p. 37). However, in the posthumous second edition of the *Principles* the words 'labour and capital' were replaced by 'expenditure', so we cannot be certain about his view. See *Principles of Political Economy*, 2nd edn (London, 1836), p. 178.

18   See, e.g., *The Works and Correspondence of David Ricardo* ed. P. Sraffa (Cambridge, 1962), III, p. 287, 'Notes on Bentham', dating from 1810 and 1811. Ricardo writes of the 'decreasing power of the land to produce in proportion to the labour and capital employed on it', and thus Sraffa says that Ricardo was already aware then of the 'principle of diminishing returns on land' (IV, p. 7). See also Ricardo to Malthus, 18 December 1814 (IV, p. 162).

19   It appears from a letter to Malthus of 6 February 1815 that he had not started writing this yet, for he says that 'I will try if I have a little leisure to put my thoughts on this subject on paper' (Sraffa, op. cit., VI, p. 173). Parts of the correspondence with Malthus reappear in his *Essay*, published on 24 February 1815.

20   Sraffa, VI, p. 179.

21   Ibid., pp. 18–41.

22   J. Mill to Ricardo, 10 October 1815. Sraffa, VI, p. 309.

23   D. Ricardo, *On the Principles of Political Economy and Taxation* (London, 1817).

24   Ricardo to J. B. Say, 18 August 1815 (Sraffa, VI, p. 249).

25   Ricardo, *Principles*, p. 56. See also S. Hollander, 'Some Technological Relationships in "The Wealth of Nations" and Ricardo's "Principles",' *CJEPS*, 1966, 184–201 and D. Simpson, 'Further Technological Relationships in "The Wealth of Nations" and in Ricardo's "Principles",' *CJEPS*, 1967, 585–90.

26   This letter of Barton's to Ricardo is presumably lost, but Ricardo's reply has been preserved (see Sraffa, VII, pp. 155–9). Later in 1817 Barton published his ideas in a pamphlet, *Observations on the Circumstances which Influence the Condition of the Labouring Classes of Society* (London, 1817). See also G. Sotiroff, 'John Barton (1789–1852)', *EJ*, 1952, 87–102.

27   Ricardo, *Principles*, p. 157.

28   Sraffa, I, lviii.

29   Sraffa, VIII, p. 171.

30   T. R. Malthus, *Principles of Political Economy Considered with a view to their Practical Application*, 1st edn (London, 1820).

31   Ricardo to J. Mill, 1 January 1821 (Sraffa, VIII, p. 331).

32   Malthus, *Principles*, pp. 261, 262, 264, 402, 409, 412 and 413.

33   My italics.

34   Sraffa, II, pp. 236, 239, 251 and 365.

35   Sraffa, I, pp. 386 and 388.

36   Hicks is wrong in saying that Ricardo gives no numerical examples in this chapter. J. R. Hicks, *A Theory of Economic History* (London, 1969), p. 168.

37   J. R. Hicks, *A Theory of Economic History* (London, 1969), pp. 150–6

and 168–71. See also J. R. Hicks, 'A Neo-Austrian Growth Theory', *EJ*, 1970, 275 and 276, and the discussion between Beach and Hicks in E. F. Beach, 'Hicks on Ricardo on Machinery', *EJ*, 1971, 916–22 and Hicks' reply, ibid., 922–5. See, further, J. R. Hicks, *Capital and Time* (Oxford, 1973), pp. 97–9 and 120. An argument similar to Hicks' is found in J. E. Tozer, 'Mathematical Investigation on the Effect of Machinery', published in the *Transactions of the Cambridge Philosophical Society*, 1828 and reprinted in 1968 by A. M. Kelley, pp. 507–20.

38 Sraffa, I, pp. 392 and 395.
39 F. A. Hayek, 'The Ricardo Effect', *Economica*, 1942, 127–52. See also C. E. Ferguson, 'The Specialization gap: Barton, Ricardo and Hollander', *HPE*, 1973, 5–11.
40 This is also stressed by S. Hollander in 'The Development of Ricardo's Position on Machinery', *HPE*, 1971, 104.
41 This hypothesis also features in Johansen's model of Classical economics, based mainly on Ricardo's ideas and incorporating technical development. L. Johansen, 'A Classical Model of Economic Growth', in *Socialism, Capitalism and Economic Growth*, essays dedicated to M. Dobb, ed. C. H. Feinstein, (Cambridge, 1967), pp. 13–29. Cf. also J. L. Cochrane, 'The First Mathematical Ricardian Model', *HPE*, 1970, 419–31.
42 Sraffa, I, pp. 396 and 397.
43 In his recent interpretation of the Ricardian model, H. Brems also incorporates technical development embodied in capital goods. H. Brems, 'Ricardo's Long-Run Equilibrium', *HPE*, 1970, 225–45. See also his book, *Labour, Capital and Growth* (London, 1973), chapter 1.
44 See the correspondence with McCulloch, beginning on 5 June 1821, in Sraffa, VIII, pp. 381ff.
45 J. C. Sismondi, *Nouveaux principes d'économie politique*, 1st edn (Paris, 1819), I, pp. 76–83; 2nd edn (Paris, 1827), I, pp. 74–82. The discussion on the equilibrium between expenditure and production, added to the second edition, is particularly important in connection with the controversy over Say's law (vol. II, pp. 369ff). See also T. Sowell, 'Sismondi: A Neglected Pioneer', *HPE*, 1972, 62–88 and Dupuigrenet-Desroussiles, 'Sismondi, la machine et le système', *Revue d'économie politique*, 1973, 334–9.
46 J. R. McCulloch, *The Principles of Political Economy; With a Sketch of the Rise and Progress of the Science* (Edinburgh, 1825), pp. 181 and 182. For a discussion of McCulloch's reaction to Ricardo's new views, see O. St. Clair, *A Key Ricardo* (New York, 1965), pp. 232ff.
47 J. R. McCulloch, *The Literature of Political Economy, A Classified Catalogue* (London, 1845), p. 286; see also D. P. O'Brien and *J. R. McCulloch, A Study in Classical Economics* (London, 1970), pp. 302–6.
48 K. Marx, *Capital, A Critical Analysis of Capitalist Production*, ed. F. Engels (Lawrence and Wishart edn, London, 1974), I, p. 202.
49 Ibid., pp. 405, 413, 574 and 583.
50 A survey is given by T. Sowell, *Say's Law, An Historical Analysis* (Princeton, New Jersey, 1972).

51 J. Mill *Commerce Defended* (London, 1807). Cf. also D. N. Winch, *James Mill, Selected Economic Writings* (London, 1966), p. 135.

52 See, e.g., J. S. Mill, *Essays on Some Unsettled Questions of Political Economy* (London, 1844), pp. 47–75.

53 J. B. Say, , *Traité d'économie politique, ou simple exposition de la manière dont se forment, se distribuent et se consomment les richesses*, 1st edn (Paris, 1803), vol. 1, pp. 152–5.

54 Ricardo, *Principles*, p. 400 and Sraffa, I, p. 290. Cf. also J. B. Say, op. cit., p. 153: '... ce n'est point tant l'abondance de l'argent qui rend les débouchés faciles, que l'abondance des autres produits en général' ('it is not so much the abundance of money as the abundance of other products in general that makes the disposal of goods easy').

55 See, e.g., M. Blaug, *Economic Theory in Retrospect* (Homewood, 1962), p. 137.

56 T. R. Malthus, *Principles of Political Economy* (London, 1820), pp. 353ff., and Sraffa, II, pp. 303ff.

57 J. S. Mill, *Essays on Some Unsettled Questions of Political Economy* (London, 1844), pp. 47–75.

58 Cf. also Ricardo to Malthus, 24 January 1817: '... I put these immediate and temporary effects quite aside, and fix my whole attention on the permanent state of things which will result from them' (Sraffa, VII, p. 120).

59 J. B. Say, *Cours complet d'économie politique pratique* (Paris, 1828), vol. 2, p. 299.

60 Say, *Traité*, 1st edn, pp. 46 and 53.

61 Say, *Cours*, I, p. 397.

62 Say, *Traité*, 2nd edn (Paris, 1814), p. 54. The chapter in question was republished unchanged in the third edition (1817).

63 Sraffa, VIII, p. 387.

64 M. Blaug, *Ricardian Economics* (New Haven, 1958), p. 74. For this contradiction in general, see also A. Kähler, *Die Theorie der Arbeitfreisetzung durch die Maschine*, (Kiel, 1933), pp. 14–35.

65 A. Kruse, *Technischer Fortschritt und Arbeitslosigkeit* (Munich, 1936), p. 14.

66 E. Lederer, *Technischer Fortschritt und Arbeitslosigkeit* (Tübingen, 1931), pp. 3 and 4.

67 J. S. Mill, *Principles of Political Economy with some of their Applications to Social Philosophy*, vol. 1 (London, 1848), p. 114.

68 On the detailed role of technology in Mill's theory see S. Hollander, 'Technology and Aggregate Demand in J. S. Mill's Economic System', *CJEPS*, 1964, 175–84.

69 The first edition of C. Babbage's book *On the Economy of Machinery and Manufactures* was published in London at the beginning of 1832 and a second edition appeared towards the end of the same year; a third, extended edition came out in the following year, erroneously giving 1832 as the year of publication. It is this third edition that will be discussed below.

70 No account is taken here of the artificial device of eliminating qualitative

differences by the use of efficiency units. Cf. J. Robinson, *The Economics of Imperfect Competition* (London, 1950), pp. 343–5 (1st edn, London, 1933).

71 The connection between this hypothesis and the wage fund is elucidated in S. Hollander, 'The Role of Fixed Technical Coefficients in the Evolution of the Wages-Fund Controversy, *OEP*, 1968, 320–41.

72 Mention should also be made here of Say's treatment of the entrepreneur, to whom he paid more attention than did the other Classical economists. See, especially, G. Koolman, 'Say's Conception of the Role of the Entrepreneur', *Economica*, 1971, 269–86.

73 The possibilities arising when labour is replaced by machines may also explain why Nassau William Senior came to more optimistic conclusions than Ricardo. See N. W. Senior, *Three Lectures on the Rate of Wages* (London, 1830), pp. 40ff. and *An Outline of the Science of Political Economy* (London, 1836), pp. 162ff.

74 K. Wicksell, *Lectures on Political Economy*, vol. 1 (London, 1935), p. 138.

75 J. S. Mill, *Principles of Political Economy*, I, p. 114 (my italics).

76 Ibid., p. 310.

CHAPTER 3: MARX

1 J. A. Schumpeter, *Capitalism, Socialism and Democracy*, 3rd edn (New York, 1950), p. 22. For the connections between Marx and the British Classical School, see also A. Walter, 'Karl Marx, the Declining Rate of Profit and British Political Economy', *Economica*, 1971, 362–77.

2 C. Wright Mills, *The Marxists* (Harmondsworth, 1969), p. 42.

3 See H. J. Sherman, 'Marxist Models of Cyclical Growth', *HPE*, 1971, 28–55; S. Maital, 'Is Marxian Growth Crisis-Ridden?', *HPE*, 1972, 113–26; D. J. Harris, 'On Marx's Scheme of Reproduction and Accumulation', *JPE*, 1972, 505–22; R. V. Eagly, 'A Macro Model of the Endogenous Business Cycle in Marxist Analysis', *JPE*, 1972, 523–39.

4 See L. R. Klein, 'Theories of Effective Demand and Employment', *JPE*, 1947, included in *Marx and Modern Economics* (London, 1968), pp. 156ff.

5 K. Marx, *Capital—A Critical Analysis of Capitalist Production*, ed. F. Engels (Lawrence and Wishart edn, London, 1974), vol. 1, p. 543.

6 B. Higgins, *Economic Development* (London, 1959), p. 109.

7 M. Bronfenbrenner, 'Eine makroökonomische Auffassung von Marx' 'Kapital', *Jahrbücher für Nationalökonomie und Statistik*, 1969, 347ff. Cf. also Peter Erdös, 'The Application of Marx's Model of Expanded Reproduction to the Trade Cycle Theory', in *Socialism, Capitalism and Economic Growth*, essays dedicated to M. Dobb and edited by C. H. Feinstein (Cambridge, 1967), pp. 59ff.

8 Marx, *Capital*, vol. 1, pp. 329, 339, 347, 348, 363, 396, 398, 575, 579, 580 and 581.

9 K. Marx, *Grundrisse der Kritik der politischen Ökonomie* (Berlin, 1953), p. 293.

10 Marx, *Capital*, vol. 1, pp. 583, 586, 588 and 589.

11 Ibid., vol. 3, p. 103.

12 Ibid., vol. 1, pp. 419, 420, 427 and 588.

13 P. A. Samuelson, 'Wages and Interest: A Modern Dissection of Marxian Economic Models', *AER*, 1957, 884ff., reprinted in *The Collected Scientific Papers of Paul A. Samuelson*, vol. 1, ed. J. E. Stiglitz (Cambridge, Mass., 1966), pp. 341ff. For Samuelson's interpretation of Marx, see also his article, 'Understanding the Marxian Notion of Exploitation: A Summary of the So-Called Transformation Problem Between Marxian Values and Competitive Prices', *JEL*, 1971, 399–431.

14 M. Blaug, 'Technical Change and Marxian Economics', *Kyklos*, 1960, 495–509, reprinted in *Marx and Modern Economics*, p. 229.

15 See, e.g., J. Robinson, 'Notes on Marx and Marshall', *Collected Economic Papers*, vol. 2 (London, 1964), p. 21.

16 Samuelson in *AER*, op. cit., p. 351.

17 Samuelson in *AER*, op. cit., pp. 359 and 360.

18 For a generalization see C. C. von Weizsäcker, 'Modern Capital Theory and the Concept of Exploitation', *Kyklos*, 1973, pp. 249–51.

19 M. Morishima, *Equilibrium, Stability and Growth* (Oxford, 1964), pp. 136ff. and M. Morishima, *Theory of Economic Growth* (Oxford, 1969), pp. 102ff. See also M. Morishima, *Marx's Economics*, (Cambridge, 1973), which is discussed by C. C. von Weizsäcker in 'Morishima on Marx', *EJ*, 1973, 1245–54; K. Förstner, 'Wirtschaftlicher Wachstum bei vollständiger Konkurrenz und linearer Technologie', in *Operations Research Verfahren*, ed. R. Henn (Meisenheim am Glan, 1967), pp. 207ff.; and M. Morishima, 'Marx's Economics: A Comment on C. C. von Weizsäcker's Article', *EJ*, 1974, pp. 387–91.

20 J. A. Schumpeter, *Capitalism, Socialism and Democracy*, pp. 31 and 32.

21 Corresponding conclusions can be obtained if *m* is not zero.

22 Joan Robinson also believes that, in Marx's analysis, '. . . technical progress normally takes forms which raise the ratio of capital to output'. 'The Model, of an Expanding Economy' in *Collected Papers*, vol. 2 (London, 1964), p. 83, originally published in *EJ*, 1962, 42ff.

23 However, see his statement: 'Aber auf einem gewissen Höhepunkt der Industrie muss die Disproportion abnehmen, dass heisst die Produktivität der Agrikultur sich relativ rascher vermehren als die der Industrie.' ('Once industry has reached a certain peak, however, the disproportionality must decrease, i.e. the productivity must increase faster in agriculture than in industry.'), *Theorien über den Mehrwert*, 5th edn (Berlin, 1923), vol. 2, p. 280.

24 J. Steindl, 'Karl Marx and the Accumulation of Capital', in *Marx and Modern Economics*, p. 252.

25 F. M. Gottheil, 'Increasing Misery of the Proletariat; an Analysis of Marx's Wage and Employment Theory', *CJEPS*, 1962, 103ff.

26 M. Blaug in *Marx and Modern Economics*, p. 242.

27 J. R. Hicks, *Theory of Wages*, 2nd edn (London, 1964), pp. 121ff.

28 A. C. Pigou, *The Economics of Welfare*, 4th edn (London, 1960), pp. 671–6.

29 R. F. Harrod, *Towards a Dynamic Economics* (London, 1956), p. 23.

30 See A. Heertje, 'An Essay on Marxian Economics', *Schweizerische Zeitschrift für Volkswirtschaft und Statistik*, 1972, 33–45, reprinted in *The Economics of Marx*, ed. M. C. Howard and J. E. King (London, 1976), pp. 219–32. See also the discussion of my exposition by the editors in the introduction to this volume and in their book *The Political Economy of Marx*, (Harlow, 1975).

31 This is also assumed by E. Wolfstetter, 'Surplus Labour and Synchronised Labour Costs in Marx's Labour Theory of Value', *EJ*, 1973, 787–809; see also I. Steedman, 'Positive Profits with Negative Surplus Value', *EJ*, 1975, 114–23, and B. Schefold, 'Wert und Preis in der marxistischen und neo-Keynesianischen Akkumulationstheorie', *Mehrwert*, 1974, 115–75.

32 See, e.g., N. Okishio, 'Technical Change and the Rate of Profit', *Kobe University Economic Review*, 1961, 85–90; M. Morishima, 'Marx in the Light of Modern Economic Theory', *Economica*, 1974, 611–32.

33 R. V. Eagly, *The Structure of Classical Economic Theory* (London, 1974), chapter 6.

34 P. Mattick, *Marx and Keynes* (London, 1974), p. 193.

35 Paul A. Baran and Paul M. Sweezy, *Monopoly Capital* (Harmondsworth, 1966), pp. 18 and 19.

CHAPTER 4: TECHNICAL DEVELOPMENT UP TO 1900

1 Extensive use is made here of *Technology in Western Civilization*, vol. 1 (New York, 1967), ed. M. Kranzberg and C. W. Pursell Jr.

2 Ibid., p. 45.

3 Ibid., p. 58.

4 Ibid., p. 101.

5 See John Davies, *A Collection of the Most Important Cases respecting Patents of Invention* (London, 1816), pp. 36ff.

6 As can be seen from a study by D. C. North on shipping in the period 1600–1850, a rise in productivity is not always due to technical development, for here it was due to a large extent to a decline in piracy and improved organization. D. C. North, 'Sources of Productivity Change in Ocean Shipping, 1600–1850', *JPE*, 1968, 953–70.

7 See note 69, Chapter 2.

8 Babbage, op. cit., p. 17. See also the introduction, in three parts, to A. Ure, *The Philosophy of Manufacturers* (London, 1835), p. 27.

9 Ibid., pp. 36, 37, 120 and 122.

10 For the connections between Babbage and Italy, see L. Bulferetti, 'Charles Babbage Economista e l'Italia', *Rassegna economica*, 1967, 1223–49.

11 Babbage, op. cit., pp. 202, 260, 263, 264, 265 and 267.

12 J. A. Schumpeter, *Theorie der wirtschaftlichen Entwicklung* (Leipzig, 1912), p. 178 and *Business Cycles* (New York, 1939), pp. 84–7.

13 See F. Machlup, 'Erfindung und technische Forschung' in *Handwörterbuch der Sozialwissenschaften*, vol. 3 (Tubingen, 1961), pp. 281 and 282, and F. Machlup, 'The Supply of Inventors and Invention', *Weltwirtschaftliches Archiv*, 1960, 210–54.

14 See E. Robinson and A. E. Musson, *James Watt and the Steam Revolution* (New York, 1969).

15 On Boulton's importance, see an article by his great-grandson: P. M. Thomas, 'Matthew Boulton—Eighteenth-century Entrepreneur par excellence', *The Philosophical Journal*, 1973, 55–70.

16 E. S. Ferguson, 'The Steam Engine before 1830', in *Technology in Western Civilization*, p. 254.

17 S. Carnot, *Réflexions sur la puissance motrice du feu et sur les machines propres à déveloper cette puissance* (Paris, 1824).

18 See A. Birch, *The Economic History of the British Iron and Steel Industry, 1784–1879*, (London, 1967).

19 Ibid., p. 325.

20 See also I. M. Tarbell, *The Life of Elbert H. Gary, A Story of Steel* (New York, 1926).

21 B. Dibner, 'The Beginning of Electricity' in *Technology in Western Civilization*, p. 437.

22 H. I. Sharlin, 'Applications of Electricity' in *Technology in Western Civilization*, p. 568.

23 Marx, *Capital*, vol. 1, chapter 15.

24 Ibid., pp. 357 and 358.

25 Ferguson, op. cit., p. 258.

26 Marx, op. cit., p. 365.

CHAPTER 5: THE BEGINNINGS OF PRODUCTION THEORY

1 W. S. Jevons, *Theory of Political Economy* (London, 1871), p. 266.

2 A. Walras, *De la nature de la richesse et de l'origine de la valeur* (Paris, 1831) and *Théorie de la richesse sociale* (Paris, 1849).

3 H. H. Gossen, *Entwicklung der Gesetze des menschlichen Verkehrs und der daraus fliessenden Regeln für menschliches Handeln* (Berlin, 1854). The 1889 edition is recommended.

4 A. A. Cournot, *Souvenirs*, published by E. P. Bottinelli, (Paris, 1913), p. 162; cf. also A. Aupetit, 'L'oeuvre économique de Cournot', *Revue de metaphysique et de morale*, 1905, 397.

5 For this, see, for example, C. L. Fry and R. B. Ekelund Jr., 'Cournot's Demand Theory: A Reassessment', *HPE*, 1971, 190–7.

6 A. A. Cournot, *Recherches sur les principes mathématiques de la théorie des richesses* (Paris, 1838), pp. 46 and 65.

7 A. A. Cournot, *Principes de la théorie des richesses* (Paris, 1863) and *Revue sommaire des doctrines économiques*, (Paris, 1877).

8 *Revue sommaire*, pp. 66, 296 and 310.

9 See V. K. Dimitriev, *Economic Essays on Value, Competition and Utility*, ed. D. M. Nuti, (Cambridge, 1974). The original Russian edition was published in 1954.

10 J. Rae, *Statement of some New Principles on the Subject of Political Economy, exposing the Fallacies of the System of Free Trade, and of some other Doctrines maintained in the 'Wealth of Nations'* (Boston, 1834). This book was republished in 1905 in the care of C. W. Mixter, with the very misleading title *The Sociological Theory of Capital*. Cf. also L.

Robbins, *The Theory of Economic Development in the History of Economic Thought* (New York, 1968), pp. 50–4.

11 J. A. Schumpeter, *History of Economic Analysis* (London, 1961), p. 468.

12 E. von Böhm-Bawerk, *Geschichte und Kritik der Kapitalzins-Theorien*, 4th edn (Jena, 1921), pp. 277–317; the discussions on Rae were added to the second edition published in 1900.

13 I. Fisher, *The Theory of Interest* (New York, 1930); Fisher dedicated this work to Rae.

14 Rae, op. cit., pp. 118–20. For Rae's concept of liquidity preference, see N. Edmonson, 'John Rae and Liquidity Preference', *HPE*, 1970, 432–40.

15 Rae, op. cit., pp. 15, 31, 213, 240 and 264.

16 C. Menger, *Grundsätze der Volkswirthschaftslehre* (Vienna, 1871), pp. 26ff. Technical development is mentioned in passing in the second edition, published in 1923 (p. 96).

17 Ibid., p. 128.

18 E. Streissler is very emphatic on this point in 'To What Extent was the Austrian School Marginalist?', *HPE*, 1972, 430 and 431.

19 W. S. Jevons, *Theory of Political Economy*. On the other hand, he did have an opportunity to do this in his book *The Coal Question; An Inquiry Concerning the Progress of the Nation, and the Probable Exhaustion of our Coal Mines* (London, 1866).

20 For this see, for example, R. D. Collison Black's introduction to the Pelican edition of Jevons' *Theory of Political Economy* (London, 1970), pp. 27 and 28; see also G. J. Stigler, *Production and Distribution Theories* (New York, 1947), pp. 22–6.

21 Jevons, op. cit., p. 221.

22 In a critical article, Steedman argues that the interest is not determined, and that no significance is attached to the concept of production period in Jevons' theory. I. Steedman, 'Jevons' Theory of Capital and Interest', *The Manchester School of Economic and Social Studies*, 1972, 31–51.

23 L. Walras, *Eléments d'économie politique pure* (Lausanne, 1877).

24 'Equations de la capitalisation et du crédit', included in L. Walras, *Théorie mathématique de la richesse sociale* (Lausanne, 1883), pp. 82ff.

25 Walras, op. cit., p. 107. On this, see also D. Collard, 'Léon Walras and the Cambridge Caricature', *EJ*, 1973, 473 and 474.

26 Montgomery ignores this aspect of Walras's theory in his article which is otherwise well worth reading: W. D. Montgomery, 'An Interpretation of Walras' Theory of Capital as a Model of Economic Growth', *HPE*, 1971, 278–97.

27 J. A. Schumpeter, *History of Economic Analysis*, p. 846.

28 F. A. Lutz, *Zinstheorie*, 2nd edn (Tübingen, 1967), p. 15.

29 E. von Böhm-Bawerk, *Kapital und Kapitalzins, II, Positive Theorie des Kapitales*, 4th edn (Jena, 1921), vol. 1, pp. 111–13. The first edition had been published in 1888.

30 For an analysis of the connection between roundabout production methods and interest with the aid of a model, see P. Bernholz, 'Superiority of Roundabout Processes and Positive Rate of Interest. A

Simple Model of Capital and Growth', *Kyklos*, 1971, 687–718. Cf. also P. Bernholz and M. Faber, 'Technical Productivity of Roundabout Processes and Positive Rate of Interest. A Capital Model with Depreciation and n-Period Horizon', *Zeitschrift für die gesamte Staatswissenschaft*, 1973, 46–61.

31 E. von Böhm-Bawerk, *Kapital und Kapitalzins, II, Positive Theorie des Kapitales*, vol. 2; see also a collection edited by F. X. Weiss, *Eugen von Böhm-Bawerks kleinere Abhandlungen über Kapital und Zins* (Vienna and Leipzig, 1926), pp. 132ff.

32 Ibid., pp. 4, 6 and 22.

33 On the other hand, Cassel believes that a lengthening of the roundabout route of production is important only if it involves a change in the technique of production. G. Cassel, *On Quantitative Thinking in Economics* (Oxford, 1935), p. 22.

34 J. R. Hicks, 'Die österreichische Kapitaltheorie und ihre Wiedergeburt in der modernen Wirtschaftswissenschaft', *Zeitschrift für Nationalökonomie*, 1972, 93. For a more detailed criticism of the vagueness of the concept of mean production period, see for example G. J. Stigler, *Production and Distribution Theories* (New York, 1941), pp. 206ff.; I. Fisher, *The Theory of Interest* (New York, 1930), p. 475; and E. Schneider, *Volkswirtschaft und Betriebswirtschaft* (Tübingen, 1964), pp. 24–35.

35 Krelle believes that Böhm's approach has, rightly, receded to the background. W. Krelle, *Produktionstheorie* (Tübingen, 1967), p. 72.

36 K. Wicksell, *Über Wert, Kapital und Rente nach den neueren national-ökonomischen Theorien* (Jena, 1893), pp. 95–127. For the connection between Böhm and Wicksell, see also F. A. Lutz, *Zinstheorie*, 2nd edn (Tübingen, 1967), pp. 26ff.

37 Wicksell regards *K* as an exogenous quantity, although strictly speaking it is an endogenous one (which fits in better with Böhm's views). Cf. J. Hirshleifer, 'A Note on the Böhm-Bawerk-Wicksell Theory of Interest', *RES*, 1967, 194.

38 A similar diagram is to be found in Wicksell, op. cit., p. 97 and, for example, in E. Helmstädter, *Der Kapitalkoeffizient* (Stuttgart, 1969), p. 153.

39 See E. Helmstädter, op. cit., pp. 171–6, and N. Reetz, *Produktionsfunktion und Produktionsperiode* (Göttingen, 1971), pp. 199–259.

40 Böhm-Bawerk, op. cit., vol. 2, p. 116.

41 J. H. von Thünen, *Der isolierte Staat in Beziehung auf Landwirtschaft und Nationalökonomie*, 1st edn (Hamburg, 1826), 2nd edn, vol. 1 (Rostock, 1842), vol. 2, Part 1 (Rostock, 1850), and vol. 2, Part 2 (Rostock, 1863).

42 *Memorials of Alfred Marshall*, ed. A. C. Pigou (London, 1925), pp. 100, 360 and 412. See also H. D. Dickinson, 'Von Thünen's Economics', *EJ*, 1969, 894–902. The emphasis von Thünen puts on education is discussed in B. F. Kikev, 'Von Thünen on Human Capital', *OEP*, 1969, 339–43.

43 Von Thünen, op. cit., vol. 2, pp. 48, 54, 97–140 (particularly p. 190) and 162–76.

44 P. J. Lloyd, 'Elementary Geometric/Arithmetic Series and Early Production Theory', *JPE*, 1969, 33. Beckmann gave a modern formulation of von Thünen's theory on the basis of this Cobb-Douglas production function. M. J. Beckmann, 'Von Thünen Revisited: A Neoclassical Land Use Model', *The Swedish Journal of Economics*, 1972, 1–7. Beckmann's analysis is further elaborated by B. M. Renaud, 'On a Neoclassical Model of Land Use', *The Swedish Journal of Economics*, 1972, 400–5. A generalization of von Thünen's theory is given by W. Buhr, 'An Operational Generalized Version of Von Thünen's Model', *Zeitschrift für die gesamte Staatswissenschaft*, 1970, 427–32.

45 Von Thünen, op. cit., vol. 2, Part 1, p. 161.

46 P. H. Wicksteed, *An Essay on the Coordination of the Laws of Distribution* (London, 1894), p. 189.

47 F. Y. Edgeworth, *Papers Relating to Political Economy*, vol. 3 (London, 1925), p. 54.

48 See Stigler, op. cit., pp. 48 and 323–6, as well as W. Jaffe, 'New Light on an Old Quarrel', *Cahiers Vilfredo Pareto*, 1964, 61–102.

49 Wicksteed, op. cit., pp. 4, 8, 15, 32, 33 and 39.

50 P. H. Wicksteed, *The Common Sense of Political Economy* (London, 1910), pp. 373 and 529.

51 J. Dorfman, 'Wicksteed's Recantation of the Marginal Productivity Theory', *Economica*, 1964, 294.

52 *Correspondence of Léon Walras and Related Papers*, ed. W. Jaffé, vol. 2 (Amsterdam, 1965), pp. 619, 644 and 653.

53 Walras, *Recueil publié par la Faculté de Droit à l'occasion de l'exposition nationale* (Geneva, 1896), pp. 3–11.

54 E. Barone, *Principi di economica politica* (Rome, 1913), pp. 8–12 and 99–102.

55 A. Marshall, *Principles of Economics* (London, 1890). Whitaker describes how this book came to be written in 'Alfred Marshall: The Years 1877 to 1885', *HPE*, 1972, 1–61. See also J. K. Whitaker, 'The Marshallian System in 1881: Distribution and Growth', *EJ*, 1974, 1–17.

56 A. Marshall, *The Present Position of Economics* (London, 1885), p. 25; see also *Memorials of Alfred Marshall*, ed. A. C. Pigou (London, 1925), p. 159.

57 Marshall, op. cit., p. 315 (Guillebaud edn, London, 1961, p. 255).

58 Ibid., p. 316.

59 See also Marshall, *Industry and Trade* (London, 1919), pp. 206ff.

60 Marshall, *Principles* (Guillebaud edn), pp. 274, 276, 285, 315, 321 and 332.

61 K. Wicksell, *Lectures on Political Economy*, vol. 1 (London, 1934), pp. 111 and 128.

62 F. Y. Edgeworth, *Papers Relating to Political Economy*, vol. 3 (London, 1925), p. 54.

63 F. Y. Edgeworth, 'The Laws of Increasing and Diminishing Returns',

EJ, 1911, reprinted in *Papers Relating to Political Economy*, vol. 1 (London, 1925), pp. 61 and 68.

CHAPTER 6: SCHUMPETER

1 N. Kaldor, 'The Relation of Economic Growth and Cyclical Fluctuations', *EJ*, 1954, 53, reprinted in *Essays on Economic Stability and Growth* (London, 1960), pp. 213 and 214.

2 P. M. Sweezy, *The Theory of Capitalist Development* (New York, 1968), p. 95 (1st edn published 1942). See also E. von Beckenrath, *Lynkeus* (Tübingen, 1962), pp. 188 and 189.

3 P. A. Samuelson, 'Schumpeter as a Teacher and Economic Theorist' in *The Collected Scientific Papers of Paul A. Samuelson*, vol. 2 (Cambridge, Mass., 1966), p. 1552.

4 J. A. Schumpeter, *History of Economic Analysis* (London, 1961), p. 1035.

5 J. A. Schumpeter, *Das Wesen und der Hauptinhalt der theoretischen Nationalökonomie* (Leipzig, 1908), pp. 614–22.

6 J. A. Schumpeter, *Theorie der wirtschaflichen Entwicklung* (Leipzig, 1912; 2nd revised edn, Leipzig, 1926); the English translation by R. Opie was published under the title of *The Theory of Economic Development* (Cambridge, 1934). The 1961 edition published by Oxford University Press, New York, is recommended. See pp. 63, 64, 66, 89, 223, 228 and 231.

7 J. A. Schumpeter, *Business Cycles*, vols 1 and 2 (New York, 1939), pp. 38, 84, 86 and 87.

8 J. A. Schumpeter, *Capitalism, Socialism and Democracy* (New York, 1942); the third edition, published in 1950, is recommended. See pp. 82, 83, 110, 132 and 223.

9 J. A. Schumpeter, *History of Economic Analysis*, pp. 679 and 1026–53.

10 P. Hennipman, 'Van en over Schumpeter', *De Economist*, 1953, 292.

11 B. Higgins, *Economic Development, Principles, Problems and Policies* (London, 1959), p. 141.

12 S. Kuznets, *Economic Change* (London, 1954), p. 112.

13 Cf. also R. S. Bhambri, 'Enterprise, Initiative and Economic Policy', *Kyklos*, 1962, 409: 'Since enterprise is defined as the carrying out of new combinations, and since new combinations are the defining feature of economic development, entrepreneurship and economic development become the same thing.'

14 This view is to be found for example in I. Adelman, *Theories of Economic Growth and Development* (London, 1962), p. 100; B. Higgins, op. cit., p. 126; R. R. Nelson *et al.*, *Technology, Economic Growth and Public Policy* (Washington, 1967), p. 18.

15 C. S. Solo, 'Innovation in the Capitalist Process: A Critique of the Schumpeterian Theory', *QJE*, 1951, 425.

16 W. P. Strassmann, *Risk and Technological Innovation* (New York, 1959), p. 218; cf. also V. W. Ruttan, 'Usher and Schumpeter on Invention, Innovation and Technological Change', *QJE*, 1959, 596ff.

17 Cf., e.g., R. Rexhausen, *Der Unternehmer und die volkswirtschaftliche Entwicklung* (Berlin, 1960); E. Carlin, 'Schumpeter's Constructed Type—The Entrepreneur', *Kyklos*, 1956, 27ff.

18 O. H. Taylor, 'Schumpeter and Marx: Imperialism and Social Classes in the Schumpeterian System', *QJE*, 1951, 541.

19 M. S. Khan, *Schumpeter's Theory of Capitalist Development* (Aligarh, India, 1957) p. 97.

20 E. Mansfield, *Industrial Research and Technological Innovation, An Econometric Analysis* (New York, 1968), p. 83. Cf. also E. Mansfield, *The Economics of Technological Change* (London, 1969), pp. 99ff.

21 W. D. Nordhaus, *Invention, Growth and Welfare, A Theoretical Treatment of Technological Change* (Cambridge, Mass., 1969), pp. 16ff.

22 W. E. G. Salter, *Productivity and Technical Change* (Cambridge, 1960), p. 17.

23 J. R. Hicks, *Capital and Growth* (Oxford, 1965), p. 298.

24 P. Sylos-Labini, *Oligopoly and Technical Progress*, revised edn (Cambridge, Mass., 1969); originally published as *Oligopolio e progresso tecnico* (Milan, 1957).

25 R. V. Clemence and F. S. Doody, *The Schumpeterian System* (Cambridge, Mass., 1950).

26 M. S. Khan believes that Schumpeter's theory and Harrod's growth theory can be combined if technical change is introduced into the picture, Kahn, op. cit., p. 113. See also a recent contribution, R. R. Nelson, S. G. Winter and H. L. Schuette, 'Technical Change in an Evolutionary Model', *QJE*, 1976, 90–118.

CHAPTER 7: TECHNICAL DEVELOPMENT SINCE 1900

1 H. Ford, *My Philosophy of Industry* (London, 1929), p. 39.

2 See D. S. Landes, *The Unbound Prometheus, Technological Change and Industrial Development in Western Europe from 1750 to the Present* (Cambridge, 1969); *Technology in Western Civilization*, vol. 2; interesting information can also be found in *Computer und Angestellte*, vols 1 and 2 (Frankfurt, 1971).

3 See, for example, Colin Clark, *The Economics of 1960* (London, 1942), pp. 22–32; J. Fourastié, *Le grand espoir du XXe siècle* (Paris, 1949), pp. 38–72; S. Kuznets, *Modern Economic Growth* (New Haven and London, 1966), pp. 86–159; S. Kuznets, *Economic Growth of Nations* (Cambridge, Mass., 1971), pp. 143–99; C. Zarka, 'Intensité du progrès technique et classification des activités économiques en trois secteurs', *Revue économique*, 1965, 199–211.

4 For the effect of technical development on the agricultural sector, see S. Ishikawa, 'Technological Change in Agricultural Production and Its Impact on Agrarian Structure, A Study on the So-Called Green Revolution', *The Economic Review*, 1971, 159–65.

5 E. Crossman, 'Taxonomy of Automation: State of the Art and Prospects', included in *Manpower Aspects of Automation and Technical Change* (an OECD publication, Paris, 1966), p. 92.

6 See, for example, R. S. Rosenbloom, 'The Transfer of Military Technology to Civilian Use', in *Technology in Western Civilization*, vol. 2, pp. 601–12.

7 See G. C. Chow, 'Technological Change and the Demand for Computers', *AER*, 1967, 1117–30.

8 For a survey, see E. Crossman, op. cit., pp. 75ff. See also P. E. Sultan and P. Prasow, 'Automation: Some Classification and Measurement Problems' in *Labour and Automation* (Geneva, 1964), pp. 9–33.

9 See J. R. Bright, 'The Development of Automation' in *Technology in Western Civilization*, vol. 2, pp. 635–55.

10 J. R. Bright, *Automation and Management* (Boston, 1958). See also the Report of the Dutch Socio-Economic Council, *Automatisering*, published in 1968, pp. 10, 11 and 12.

11 On this, see for example E. Moonman (ed.), *British Computers and Industrial Innovation* (London, 1971). See also D. J. Robertson, 'Economic Effects of Technological Change', *Scottish Journal of Political Economy*, 1965, particularly 183–7.

12 Cf. *Manpower Aspects of Automation and Technical Change* (OECD, Paris, 1966), especially the contributions of H. Reinoud and F. F. M. Moll; see also the report mentioned in note 10 above and *The Employment Impact of Technological Change* (Washington, 1966).

13 Jules Verne, *Round the Moon* (London, 1963), p. 183.

14 See also O. T. Mason, *The origins of Invention: A study of Industry among Primitive Peoples* (London, 1893), p. 18.

15 J. Jewkes, D. Sawers and R. Stillerman, *The Sources of Invention*, 2nd edn (London, 1969), p. 63.

16 Cf. also W. F. Ogborn: 'It seems very difficult to anticipate inventions and their social effects'. *Recent Social Trends in the United States* (New York, 1935), p. 166.

17 Jewkes, Sawers and Stillerman, op. cit., p. 195.

18 J. Enos, 'Invention and Innovation in the Petroleum Refining Industry', in *The Rate and Direction of Inventive Activity: Economic and Social Factors* (Princeton, 1962), pp. 299–321.

19 F. Lynn, 'An Investigation of the Rate of Development and Diffusion of Technology in our Modern Industrial Society', in *The Employment Impact of Technological Change* (Washington, 1966), pp. 31–86.

20 E. Mansfield, 'Technological Change: Measurement, Determinants and Diffusion', in *The Employment Impact of Technological Change*, op. cit., pp. 129 and 130; cf. also Mansfield *Industrial Research . . . etc*, p. 129, and *The Economics of Technological Change*, pp. 131–3 and 'Technical Change and the Rate of Imitation', *Econometrica*, 1961, 741–66. See also R. R. Nelson, M. J. Peck and E. D. Kalachek, *Technology, Economic Growth and Public Policy* (Washington, 1967), pp. 97ff.

21 J. Langish, M. Gibbons, W. G. Evans and F. R. Jevons, *Wealth from Knowledge, A Study of Innovation in Industry* (London, 1972), pp. 42–9. This book confirms the conclusions drawn above about the role of individuals and science in innovation and about the time lag between inventions and their application.

22   E. Mansfield, 'International Technology Transfer: Forms, Resources, Requirements and Policies', *AER*, 1975, 372–6.

23   G. R. Hall and R. E. Johnson, 'Transfers of United States Aerospace Technology to Japan', in *The Technology Factor in International Trade*, ed. R. Vernon (New York, 1970), pp. 334–56.

24   See I. H. Siegel, 'Scientific Discovery and the Rate of Invention', in *The Rate and Direction of Invention Activity: Economic and Social Factors* (Princeton, 1962), p. 445 and A. P. Usher, *A History of Mechanical Inventions*, 2nd edn (New York, 1954). The views of Schumpeter and Usher are compared in V. W. Ruttan, 'Usher and Schumpeter on Invention, Innovation and Technological Change', *QJE*, 1959, reprinted in *The Economics of Technological Change*, ed. N. Rosenberg (Harmondsworth, 1971), pp. 73–85. Ruttan mainly agrees with Usher. See, however, P. R. Schweitzer, 'Usher and Schumpeter on Invention, Innovation and Technological Change: Comment', *QJE*, 1961, 152–4, together with Ruttans's reply to it.

25   For the importance of the computer for the development of economic science see W. J. Frazer Jr, 'An Assessment of the Impact of the Computer', *Schweizerische Zeitschrift für Volkswirtschaft und Statistik*, 1973, 579–95.

CHAPTER 8: MICROECONOMIC PRODUCTION THEORY

1   T. C. Koopmans, *Three Essays on the State of Economic Science* (New York, 1957), p. 126.

2   See Section 5.9.

3   P. A. Samuelson in *Monopolistic Competition Theory: Studies in Impact, Essays in Honor of Edward H. Chamberlin*, ed. R. E. Kuenne (New York, 1967), pp. 109–13.

4   The most important publications in this connection are: J. H. Clapham, 'On Empty Economic Boxes', *EJ*, 1922, 305–14; D. H. Robertson, 'Those Empty Boxes', *EJ*, 1924, 16–30; P. Sraffa, The Laws of Returns under Competitive Conditions', *EJ*, 1926, 535–50; J. Viner, 'Cost Curves and Supply Curves', *Zeitschrift für Nationalökonomie*, 1931, 23–46; J. Robinson, 'Rising Supply Price', *Econometrica*, 1941, 1–8. All these were reprinted in *Readings in Price Theory*, ed. G. J. Stigler and K. E. Boulding (Chicago, 1952).

5   In *The Years of High Theory* (Cambridge, 1967), G. Shackle discusses in detail the work of Sraffa, J. Robinson and E. H. Chamberlin and is well worth reading for the background of this development.

6   H. von Stackelberg, *Grundlagen einer reinen Kostentheorie* (Vienna, 1932). Mention should also be made of A. L. Bowley, *The Mathematical Groundwork of Economics* (Oxford, 1924), especially chapter 3.

7   E. Schneider, *Theorie der Produktion* (Vienna, 1934).

8   S. Carlson, *A Study on the Pure Theory of Production* (London, 1939).

9   For this see also F. A. and V. Lutz, *The Theory of Investment of the Firm* (Princeton, 1951), p. 5.

10   C. W. Cobb and P. H. Douglas, 'A Theory of Production', *AER*, 1928, 139–65; cf. also P. H. Douglas, 'Some New Material on the Theory of

Distribution', in *Economic Essays in Honour of Gustav Cassel* (London, 1933), pp. 105–15; P. H. Douglas, *The Theory of Wages* (London, 1934).

11 J. Tinbergen, 'Zur Theorie der langfristigen Wirtschaftsentwicklung', *Weltwirtschaftliches Archiv*, 1942, 511–49, published in English in Jan Tinbergen, *Selected Papers* (Amsterdam, 1959), pp. 182–221.

12 J. R. Hicks, *Capital and Growth*, p. 299.

13 Detailed accounts of the production theory are given in R. Frisch, *Theory of Production* (Dordrecht, 1965); S. Danø, *Industrial Production Models, A Theoretical Study* (Vienna, 1966) (both these books are discussed in E. Schneider, 'Zwei Standardwerke zur Theorie der Produktion', *Weltwirtschaftliches Archiv*, 1966, vol. 2, 51–6); W. Krelle, *Produktionstheorie* (Tübingen, 1969); C. E. Ferguson, *The Neoclassical Theory of Production and Distribution* (Cambridge, 1969). A simple introduction to this subject is to be found in D. F. Heathfield, *Production Functions* (London, 1971).

14 On this subject, see O. Opitz, 'Zum technischen Optimierungsproblem des Unternehmers', *Schweizerische Zeitschrift für Vokswirtschaft und Statistik*, 1970, 369–81.

15 For a synthesis see M. Frankel, 'The Production Function in Allocation and Growth', *AER*, 1962, 995–1022.

16 A good survey of some of the statistical problems concerning the Cobb-Douglas production function is to be found in R. K. Diwan, 'On the Cobb-Douglas Production Function', *SEJ*, 1967/68, 410–14.

17 See also M. Brown, *On the Theory and Measurement of Technological Change* (Cambridge, 1966), pp. 39–42.

18 See K. J. Arrow, H. B. Chenery, B. S. Minhas and R. M. Solow, 'Capital-Labour Substitution and Economic Efficiency', *RES*, 1961, 225–50 and M. Brown and J. S. de Cani, 'Technological Change and the Distribution of Income', *IER*, 1963, 289–309; Uzawa has generalized the CES functions to the case of n factors of production: H. Uzawa, 'Production Functions with Constant Elasticities of Substitutions', *RES*, 1962, 291–9. See also V. Mukerji, 'Generalized SMAC Function with Constant Ratios of Elasticities of Substitution', *RES*, 1963, 233–61; D. McFadden, 'Constant Elasticity of Substitution Production Functions', *RES*, 1963, 73–83; H. Frisch, 'Die C.E.S.-Function', *Zeitschrift für Nationalökonomie*, 1964, 419–44; W. Scheper, 'Produktionsfunktionen mit konstanten Substitutions-elastizitäten', *Jahrbücher für Nationalökonomie und Statistik*, 1965, 1–21; P. B. Kenen, 'Efficiency Differences and Factor Intensities in the C.E.S. Production Function', *JPE*, 1966, 635 and 636; R. Sato, 'A Two-level Constant-Elasticity of Substitution Production Function', *RES*, 1967, 201–18; F. de Jesus, 'A Direct Derivation of the h-homogeneous CES Production Function: A Note', *Metroeconomica*, 1968, 294–8. Whitaker has pointed out that the CES function is implied in an article by Dickinson: H. D. Dickinson, 'A Note on Dynamic Economics', *RES*, 1955, 169–79 and J. K. Whitaker, 'A Note on the CES Production Function', *RES*, 1964, 166 and 167.

19 See also E. Helmstädter, 'Die Isoquanten gesamtwirtschaftlicher

Produktionsfunktionen mit konstanten Substitutionselastizität', *Jahrbücher für Nationalökonomie*, 1964, 177–95; T. Supel and G. Sher, 'A Note on the Asymptotes of the C.E.S. Production Function in the Case where $\sigma < 1$', *RES*, 1970, 337.

20   W. Leontief, *The Structure of the American Economy, 1919–1939*, 2nd edn (New York, 1951); see also *Studies in the Structure of the American Economy*, ed. W. Leontief (New York, 1953).

21   Some statistical problems of estimation have been pointed out by Z. S. Wurtele in 'A Problem Encountered in the Comparison of Technical Coefficients', *RES*, 1958/59, 148–52.

22   G. Tintner has therefore pointed out that the production period should be treated as a purely microeconomic concept. G. Tintner, 'Linear Economics and the Böhm-Bawerk Period of Production', *QJE*, 1974, 127–32.

23   As long ago as 1939 Hicks raised the question of the weights to be used in the calculation of the mean. *Value and Capital*, 2nd edn (Oxford, 1956), p. 218.

24   The concept may still be suitable for other purposes, as can be seen, for example, from an analysis by C. C. von Weizsäcker in 'Die zeitliche Struktur des Produktionsprozess und das Problem der Einkommensverteilung zwischen Kapital und Arbeit', *Weltwirtschaftliches Archiv*, 1971, 1–33. See also his book, *Steady State Capital Theory* (Heidelberg, 1971).

25   G. L. S. Shackle, *Epistemics and Economics, A Critique of Economic Doctrines* (Cambridge, 1972), p. 316.

26   Sraffa draws a similar conclusion from a description of a similar situation; P. Sraffa, *Production of Commodities by Means of Commodities, Prelude to a Critique of Economic Theory* (Cambridge, 1963), p. 38.

27   In Allais' study the average production period is not a purely technical concept, but is expressed explicitly in terms of value; M. Allais, 'The Influence of the Capital-Output Ratio on Real National Income', *Econometrica*, 1962, 704ff.

28   J. R. Hicks, 'The Austrian Theory of Capital and its Rebirth in Modern Economics', in *Carl Menger and the Austrian School of Economics*, ed. J. R. Hicks and W. Weber (Oxford, 1973), p. 194; see also J. R. Hicks, *Capital and Time*, pp. 14–26.

29   Thus Furono speaks of a production period although he deals with a dynamic production function. Y. Furono, 'The Period of Production in Two-Sector Models of Economic Growth', *IER*, 1965, 240–4.

30   W. Salter, *Productivity and Technical Change*, 2nd edn (Cambridge, 1966), p. 19.

31   This is what is known in econometrics as the identification problem. See J. S. Cramer, *Empirical Econometrics* (Amsterdam, 1969), pp. 118–28; A. A. Walters, *An Introduction to Econometrics* (London, 1968), pp. 163–76; and a pioneering article by T. C. Koopmans, 'Identification Problems in Economic Model Construction', *Econometrica*, 1949, 125–44.

32 An article along the same lines is W. Eichhorn, 'Deduktion der Ertragsgesetze aus Prämissen', *Zeitschrift für Nationalökonomie*, 1968, 191–205.

33 C. R. Frank Jr, *Production Theory and Invisible Commodities* (Princeton, 1969).

34 D. W. Jorgenson and L. J. Lau, 'The Duality of Technology and Economic Behaviour', *RES*, 1974, 181–200.

35 The same theory can be constructed by using non-negative numbers only, but the version described here is given preference in this book.

36 For detailed studies see T. C. Koopmans, 'Analysis of Production as an Efficient Combination of Activities' in *Activity Analysis of Production and Allocation* (New York, 1951), pp. 33–97; idem, *Three Essays on the State of Economic Science* (New York, 1957); G. Debreu, *Theory of Value* (New York, 1959); W. Wittmann, *Produktionstheorie* (Berlin, 1968). A harbinger of activity analysis is Remak. On this see W. Wittmann, 'Die extremale Wirtschaft', *Jahrbücher für Nationalökonomie und Statistik*, 1967, 397–409.

37 In the last two cases, $v (= 0)$ is ignored.

38 Cf. also R. E. Kuenne's explanation of Koopmans' production model in *The Theory of General Economic Equilibrium* (Princeton, 1963), pp. 390–4. Also of interest is Y. Otani, 'Neo-Classical Technology Sets and Properties of Production Possibility Sets', *Econometrica*, 1973, 667–82.

39 See, for example, O. Opitz, 'Zum technischen Optimierungsproblem des Unternehmers', *Schweizerische Zeitschrift für Volkswirtschaft und Statistik*, 1970, 369–81. Also of interest is O. Opitz, 'Zum Problem der Aktivitatsanalyse', *Zeitscrift für die gesamte Staatswissenschaft*, 1971, 238–55, where the theory is extended to convex technologies.

40 For some other aspects of the production function, particularly in connection with the derivation of the long-term cost function, see E. G. Furubotn, 'Long-Run Analysis and the Form of the Production Function', *Economia Internationale*, 1970, 1–33.

41 As mentioned before, Schumpeter associates innovation with the setting-up of a new production function; see Section 6.3.

42 An axiomatic representation based on this concept of technical change can be found in G. Debreu, 'Numerical Representations of Technological Change', *Metroeconomica*, 1954, 45–54.

43 See P. A. David, *Technical Choice, Innovation and Economic Growth* (Cambridge, 1975), p. 2.

44 Atkinson and Stiglitz's view that technical change is localized in a limited part of the production function and therefore does not cause a shift in the whole function is consistent with the attempt, developed here, to establish a connection with the axiomatic production theory; A. B. Atkinson and J. E. Stiglitz, 'A New View of Technological Change', *EJ*, 1969, 573–8. An elaboration, about the transfer of technical knowledge to developing countries, is given by H. Lapan and P. Bardhan, 'Localized Technical Progress and Transfer of Technology and Economic Development', *JET*, 1973, 585–95.

45 M. Brown, *On the Theory and Measurement of Technological Change*, pp. 12ff.

46 G. Schätzle, *Technischer Fortschritt und Produktionsfunktion* (Cologne, 1966), also included in the collection *Produktionstheorie und Produktionsplanung*, dedicated to K. Hax (Cologne, 1966), pp. 41–62.

47 Special difficulties arise when the firm makes not one but several products; on this see Y. Mundlak, 'Transcendental Multiproduct Production Functions', *IER*, 1964, 273–84. A study on this subject is R. E. Hall, 'The Specification of Technology with Several Kinds of Output', *JPE*, 1973, 876–92.

48 For this view, see W. Salter, *Productivity and Technical Change*, p. 14.

49 C. E. Ferguson, *The Neoclassical Theory of Production and Distribution*, p. 216.

CHAPTER 9: MACROECONOMIC PRODUCTION THEORY

1 F. Machlup, *Essays on Economic Semantics* (Englewood Cliffs, New York, 1963), p. 142.

2 See also ibid., pp. 97ff.

3 See in this connection Sen's study on the choice of the optimum production technique in a developing country: *Choice of Techniques, An Aspect of the Theory of Planned Economic Development*, 3rd edn (Oxford, 1968).

4 On this, see, for example, B. P. Stigum, 'On Certain Problems of Aggregation', *IER*, 1967, 349–68 and C. Copper (ed.), *Science, Technology and Development* (London, 1973).

5 K. Sato, *Production Functions and Aggregation* (Amsterdam, 1975).

6 J. Robinson, 'The Production Function and the Theory of Capital', *RES*, 1953/54, 81–106.

7 A. Nataf, 'Sur la possibilité de construction de certains macro-modèles', *Econometrica*, 1948, 232–44. A good survey of this kind of problem is given by H. A. J. Green, *Aggregation in Economic Analysis* (Princeton, 1964); see also B. Felderer, 'Makro-ökonomische Produktionsfunktion und optimale Betriebsgrösse', *Zeitschrift für die gesamte Staatswissenschaft*, 1972, 293–303.

8 F. M. Fisher, 'The Existence of Aggregate Production Functions', *Econometrica*, 1969, 560.

9 For the proof see R. M. Solow, *Capital, Labor and Income in Manufacturing, The Behaviour of Income Shares* (Princeton, 1964), pp. 101–28.

10 See the second half of Section 9.4 (after Figure 7).

11 J. K. Whitaker, 'Capital Aggregation and Optimality Conditions', *RES*, 1968, 429–42.

12 R. M. Solow, 'A Contribution to the Theory of Economic Growth', *QJE*, 1956, 65–94 and 'Investment and Technical Progress' in K. J. Arrow *et al.* (eds), *Mathematical Methods in the Social Sciences* (Stanford, 1960). See also P. A. Diamond, 'Technical Change and the Measurement of Capital and Output', *RES*, 1965, 289–98.

13 P. A. Samuelson, 'Parable and Realism in Capital Theory: The Surrogate Production Function', *RES*, 1962, 193–206, reprinted in *The Collected Scientific Papers of Paul A. Samuelson*, vol. 1, pp. 325ff.

14 A rather neglected but excellent account of the return of techniques can be found in D. M. Bensusan-Butt, *On Economic Growth: An Essay in Pure Theory* (Oxford, 1960), pp. 122–8. See also G. C. Harcourt, 'Some Cambridge Controversies in the Theory of Capital', *JEL*, 1969, 369–405.

15 The condition for identical ratios between the factors was somewhat weakened in J. McIntosh, 'Some Notes on the Surrogate Production Function', *RES*, 1972, pp. 505–10.

16 See also A. Bhaduri, 'On the Significance of Recent Controversies on Capital Theory: A Marxian View', *EJ*, 1969, 537. This author has also illustrated the return of techniques in an intertemporal analysis in 'A Physical Analogue of the Reswitching Problem', *OEP*, 1970, 148–55.

17 J. Robinson, *The Accumulation of Capital* (London, 1956), pp. 109 and 110.

18 pp. 81–7.

19 See, e.g., J. Robinson and K. A. Nagvi, 'The Badly Behaved Production Function', *QJE*, 1967, 578–91; J. Robinson, 'Capital Theory up to Date', *CJEPS*, 1970, 309–17; J. Robinson, 'The Measurement of Capital; The End of the Controversy', *EJ*, 1971, 597–602.

20 D. Levhari, 'A Non-Substitution Theorem and Switching of Techniques', *QJE*, 1965, 98–105.

21 L. L. Pasinetti, 'Changes in the Rate of Profit and Switches of Techniques', *QJE*, 1966, 503–17; see also the articles of M. Morishima, P. Garegnani and M. Bruno, E. Burmeister and E. Sheshinski on the switching of techniques in the 1966 volume of this Journal.

22 P. A. Samuelson and D. Levhari, 'The Nonswitching Theorem is False', *QJE*, 1966, 518–19. For a discussion of the maximum number of switch points, see K. Bharadwaj, 'On the Maximum Number of Switches Between Two Production Systems', *Schweizerische Zeitschrift für Volkswirtschaft und Statistik*, 1970, 409–29.

23 P. A. Samuelson, 'A Summing Up', *QJE*, 1966, 568. See also E. Burmeister, 'Neo-Austrian and Alternative Approaches to Capital Theory', *JEL*, 1974, 413–56.

24 For this see K. Wicksell, *Lectures on Political Economy*, vol. 1 (London, 1934), pp. 178–80; C. E. Ferguson and D. L. Hooks, 'The Wicksell Effects in Wicksell and in Modern Capital Theory', *HPE*, 1971, 353–72; also T. K. Rymes, *On Concepts of Capital and Technical Change* (Cambridge, 1971), pp. 78ff. and B. Sandelin, 'The Wicksell Effect, Dewey and Others: A Note', *HPE*, 1975, 123–32.

25 Hicks in particular refers to this in his book *Capital and Growth*, pp. 160ff. In this connection see also P. Garegnani, 'Heterogeneous Capital, the Production Function and the Theory of Distribution', *RES*, 1970, 407–36. On Hicks' contribution see especially R. P. Zuidema, 'De controverse aangaade heterogene kapitalaalgoederen' ('The Controversy Concerning Heterogeneous Capital Goods'), *De*

Economist, 1972, 401–38, which, in my view, somewhat underestimates Sraffa's contribution. An elaboration of Hicks' analysis can be found in M. Brown, 'Substitution-Composition Effects, Capital Intensity, Uniqueness and Growth', *EJ*, 1969, 334–47.

26   H. Y. Wan Jr, *Economic Growth* (New York, 1971), p. 114.

27   L. Spaventa, 'Rate of Profit, Rate of Growth and Capital Intensity in a Simple Production Model', *OEP*, 1970, 129–47.

28   Y. Ayküz, 'Income Distribution, Value of Capital and Two Notions of the Wage-Profit Trade-Off', *OEP*, 1972, 156–65.

29   K. Jaeger, 'Income Distribution, Value of Capital and Two Notions of the Wage-Profit Trade-Off', *OEP*, 1973, 286–8. Another article of interest is D. J. Harris, 'Capital, Distribution and the Aggregate Production Function', *AER*, 1973, 100–13.

30   See, for example, R. Britto, 'Some Recent Developments in the Theory of Economic Growth, an Interpretation', *JEL*, 1973, 1343–66; L. Gallaway and V. Shukla, 'The Neoclassical Production Function', *AER*, 1974, 348–58; A. Asimakopoulos and G. C. Harcourt, 'Proportionality and the Neoclassical Parables', *SEJ*, 1974, 481–3; J. F. Wright, 'The Dynamics of Reswitching', *OEP*, 1975, 21–46. See also J. E. Stiglitz, 'Recurrence of Techniques in a Dynamic Economy' and L. Spaventa, 'Notes on Problems of Transition between Techniques' in *Models of Economic Growth*, ed. J. A. Mirrlees and N. H. Stern (London, 1973).

31   The opposite view has been expressed by K. Sato, 'The Technology Frontier in Capital Theory', *QJE*, 1974, 353–84. For a generalization of Sato's approach, see E. A. C. Thomas, 'On Technological Implications of the Wage-Profit Frontier', *JET*, 1975, 263–82.

32   See D. A. Starrett, 'Switching and Reswitching in a General Production Model', *QJE*, 1969, 683. See also M. Bruno, E. Burmeister and E. Sheshinski, 'Switching Techniques and Consumption per head; an Economic Clarification', *QJE*, 1970, 533–5; C. F. Ferguson and R. F. Allen, 'Factor Prices, Commodity Prices and Switches of Technique', *Eastern Economic Journal*, 1970, 95–109.

33   M. Blaug, *The Cambridge Revolution, Success or Failure* (London, 1974).

34   Recent books on economic theory based on this insight are: J. Robinson and J. Eatwell, *An Introduction to Modern Economics* (London, 1973) and J. A. Kregel, *The Reconstruction of Political Economy: An Introduction to Post-Keynesian Economics* (London, 1973).

35   J. Robinson, 'The Unimportance of Reswitching', *QJE*, 1975, 32–9; see also L. L. Pasinetti, *Growth and Income Distribution, Essays in Economic Theory* (London, 1974).

36   Cf., for a slightly different view, H. Lydall, 'A Theory of Distribution and Growth with Economies of Scale', *EJ*, 1971, 92–3.

37   We cannot discuss here the important econometric question of whether aggregation is necessary for estimating the parameters. For this see, for example, W. D. Fisher, 'Optimal Aggregation in Multi-Equation Prediction Models', *Econometrica*, 1962, 744–69ff.; J. B. Edwards and

G. H. Orcutt, 'Should Aggregation Prior to Estimation be the Rule?', *RE and S*, 1969, 409–20.

38   G. Debreu, *Theory of Value, An Axiomatic Analysis of Economic Equilibrium* (New York, 1959), pp. 28 and 41.

39   Cf. also K. Jaeger, 'Einige Bemerkungen zur Reswitching-Diskussion', in *Probleme der Wachstumstheorie*, ed. B. Gahlen and A. E. Ott (Tübingen, 1971), pp. 138–75. G. C. Harcourt, *Some Cambridge Controversies in the Theory of Capital* (Cambridge, 1972).

40   For this see E. Burmeister and A. R. Dobell, *Mathematical Theories of Economic Growth* (London, 1970), pp. 70–3.

41   Cf. also E. J. Nell, 'Theories of Growth and Theories of Value', *Economic Development and Cultural Change*, 1967, 15–26; J. von Neumann and P. Sraffa, 'Deux contributions differentes à la critique de l'analyse traditionelle des prix et de la production', *Revue d'Economie Politique*, 1974, 872–90; A. Delarue, 'Eléments d'économie néoricardienne', *Revue Economique*, 1975, 177–9 and 337–64.

42   Sraffa, *Production of Commodities*, p. 93.

43   See, on the same lines, Yew-Kwang Ng, 'Harcourt's Survey of Capital Theory', *The Economic Record*, 1974, 121–2.

44   See, on the same lines, C. E. Ferguson, 'Capital Theory Up to Date: A Comment on Mrs Robinson's Article', *CJEPS*, 1971, 250–4; Joan Robinson's reply, 'Capital Theory Up to Date: A Reply', ibid., 254–6, together with her discussion of Ferguson's book in *CJEPS*, 1970, 309–17. Ferguson developed his standpoint in 'The Current State of Capital Theory: A Tale of Two Paradigms', *SEJ*, 1972, 160–76.

45   Vosgerau has developed a model in which land plays a role as a separate factor of production; H. J. Vosgerau, 'Boden und wirtschaftlicher Wachstum', *Kyklos*, 1972, 481–500.

46   See Section 2.5 and Sraffa, *Works of Ricardo*, p. 388.

47   Cf. Last paragraph in Section 5.9, as well as Wicksell, *Lectures on Political Economy*, pp. 133–44.

48   Ibid., p. 135.

49   A. C. Pigou, *The Economics of Welfare*, 4th edn (London, 1932), p. 674.

50   J. R. Hicks, *The Theory of Wages*, 2nd edn (London, 1964), pp. 121 and 122.

51   J. Robinson, *Essays in the Theory of Employment* (London, 1937), pp. 131–6; discussed by R. F. Harrod in *EJ*, 1937, 328 and 329.

52   J. Robinson, 'The Classification of Inventions', *RES*, 1938, 139–42.

53   R. F. Harrod, *Towards a Dynamic Economics* (London, 1948), pp. 22–8.

54   Ibid., p. 26; cf. also R. F. Harrod, 'The "Neutrality" of Improvements', *EJ*, 1961, 300–4. Harrod has recently repeated his analysis in *Economic Dynamics* (London, 1973), pp. 52–7.

55   H. Uzawa, 'Neutral Inventions and Stability of Growth Equilibrium', *RES*, 1961, 117–24.

56   For some generalizations, see also E. Burmeister and A. R. Dobell, 'Disembodied Technological Change with Several Factors', *JET*, 1969, 1–8. An elegant method for incorporating both types of neutral technical change in a two-sector growth model with fixed technical

coefficients can be found in R. N. Batra, 'Hicks- and Harrod-Neutral Technical Progress and the Relative Stability of a Two-Sector Growth Model with Fixed Coefficients', *JPE*, 1970, 84–96.

57   M. Blaug expressly states that his review article does not deal with inventions but with innovations; M. Blaug, 'A Survey of the Theory of Process-Innovations', *Economica*, 1963, 13–32.

58   C. Kennedy, 'Harrod on "Neutrality" ', *EJ*, 1962, 249–50 and C. Kennedy, 'The Character of Improvements of Technical Progress', *EJ*, 1962, 899–911.

59   For this see, e.g., Asimakopoulos, 'The Definition of Neutral Inventions', *EJ*, 1963, 675–80; A. Asimakopoulos and J. C. Weldon, 'The Classifications of Technical Progress in Models of Economic Growth', *Economica*, 1963, 372–86.

60   H. Birg, 'Zu einer allgemeinen Theorie des technischen Fortschritts, Kritik der Definitionen von J. R. Hicks und R. F. Harrod', *Jahrbücher für Nationalökonomie und Statistik*, 1969, 327–46.

61   E. Helmstädter, 'Harrod und die neoklassische Wachstumstheorie', *Zeitschrift für die gesamte Staatswissenschaft*, 1965, 433–51; *idem*, 'Typen Harrod-neutralen technischen Fortschritts', *Operations Research Verfahren*, ed. R. Henn (Meisenheim am Glan, 1967), pp. 192–210; *idem*, *Der Kapitalkoeffizient*, (Stuttgart, 1969), pp. 171–87.

62   Okuguchi has demonstrated that a necessary and sufficient condition for the neutrality in Harrod's sense is:

$$\frac{\delta F}{\delta t} \left/ L\frac{\delta F}{\delta L} \right. = \alpha(t)$$

where $\alpha(t)$ is a function of time, and the production function is $F(K, L, t)$. See K. Okuguchi, 'A Note on Harrod Neutral Technical Progress', *Metroeconomica*, 1968, 50–4.

63   R. M. Solow, *Capital Theory and the Rate of Return* (Amsterdam, 1963), pp. 75ff.

64   Cf. F. H. Hahn and R. C. Matthews, 'The Theory of Economic Growth: A Survey', *EJ*, 1964, 830.

65   R. A. McCain, 'Induced Technical Progress and the Price of Capital Goods', *EJ*, 1972, 921–33. See also C. R. Dougherty, 'On the Secular Macro-Economic Consequences of Technical Progress', *EJ*, 1974, 543–65.

66   C. Kennedy, 'A Generalization of the Theory of Induced Bias in Technical Progress', *EJ*, 1973, 48–57.

67   M. J. Beckmann and R. Sato, 'Aggregate Production Functions and Types of Technical Progress; A Statistical Analysis', *AER*, 1969, 88–101; see also M. H. Beckmann and R. Sato, 'Neutral Inventions and Production Functions', *RES*, 1968, 57–66 and R. Sato and M. J. Beckmann, 'Economic Growth, Technical Progress and the Production Function', *Jahrbücher für Nationalökonomie und Statistik*, 1975, 139–42.

68   E. R. Brubaker, 'Multi-Neutral Technical Progress; Compatibilities Conditions and Consistency with some Evidence', *AER*, 1972, 997.

69 T. Nôno, 'A Classification of Neutral Technical Changes: An Application of Lie Theory', *Bulletin of Fukuoka University of Education*, 1970, 47–62 and 1971, 43–56.

70 See, on the same lines, W. W. Chang, 'The Neoclassical Theory of Technical Progress', *AER*, 1970, 922

71 N. Kaldor, 'A Model of Economic Growth', *EJ*, 1957, 596, reprinted in N. Kaldor, *Essays on Economic Stability and Growth* (London, 1968), pp. 259–300.

72 W. Krelle, *Produktionstheorie* (Tübingen, 1969), p. 201.

73 See J. Black, 'The Technical Progress Function and the Production Function', *Economica*, 1962, 166–70. See also D. G. Champerowne, 'The Stability of Kaldor's 1957 Model', *RES*, 1971, 47–62.

74 We shall ignore here the possibility that another function arises if different price relationships are assumed.

75 This also applies to later versions of Kaldor's model; cf. N. Kaldor, 'Capital Accumulation and Economic Growth', in *The Theory of Capital* (Proceedings of a Conference held by the International Economic Association), ed. F. A. Lutz and D. C. Hague (London, 1963), pp. 177–222; and N. Kaldor and J. A. Mirrlees, 'A New Model of Economic Growth', *RES*, 1962, 172–92, reprinted in *Readings in the Modern Theory of Economic Growth*, ed. J. E. Stiglitz and H. Uzawa (Cambridge, Mass., 1969), pp. 384–402. In an article which has gone unnoticed, Jossa disputes the existence of the Golden Age in the Kaldor-Mirrlees model: B. Jossa, 'Condizioni di esistenza dell "età dell'oro" nel modello di sviluppo di Kaldor e Mirrlees', *L'Industria*, 1965, 70–91.

76 See, in the same vein, C. Kennedy, who has in addition pointed out that the application of certain types of neutral inventions does not require investment (which accumulation does require by definition); cf. C. Kennedy, 'Technical Progress and Investment', *EJ*, 1961, 292–9.

77 Kaldor's models have introduced a new element in the Cambridge controversy, the central point of which is whether investment is determined by saving, as assumed by the Neoclassical School, or whether it is determined independently of saving. This part of the Cambridge controversy was introduced by L. L. Pasinetti in his 'Rate of Profit and Income Distribution in Relation to the Rate of Growth', *RES*, 1962, 267–79. Samuelson and F. Modigliani replied to this in 'The Pasinetti Paradox in Neoclassical and More General Models', *RES*, 1966, 269–301. An excellent summary is given by J. A. Kregel, *Rate of Profit, Distribution and Growth: Two Views* (London, 1971). Some empirical investigations on Kaldor's theory have been carried out by T. F. Cripps and R. J. Tarling, *Growth in Advanced Capitalist Economies 1950–1970* (Cambridge, 1973).

78 Salter, *Productivity and Technical Change*, p. 29. For a discussion of this book, see G. C. Harcourt, 'The Rate of Profits in Equilibrium Growth Models: A Review Article', *JPE*, 1973, 1261–77.

79 A. E. Ott, 'Technischer Fortschritt', in *Handwörterbuch der Sozialwissenschaften* (Stuttgart, 1959), pp. 353ff.

80 See A. Heertje, 'Zur Ottschen Klassifikation des technischen

Fortschritts', *Zeitschrift für Nationalökonomie*, 1970, 227–30. See also Ott's reply in the same issue, 'Messung und Klassifikation des teschnischen Fortschritts, Eine Antwort an A. Heertje', 231–3. Walter believes that substitution and technical change can be separated only in the case of Hicks-neutral technical change, a conclusion reached by him on the basis of Ott's scheme. H. Walter, 'Technischer Fortschritt und Faktor-Substitution', *Jahrbücher für Nationalökonomie und Statistik*, 1963, 97–114.

81   R. M. Solow, *Capital Theory and the Rate of Return* (Amsterdam, 1963), pp. 69ff. J. Robinson, 'Solow on the Rate of Return', *EJ*, 1964, 410–17, reprinted in *Collected Economic Papers*, vol. 3 (Oxford, 1965), pp. 36–47. Pasinetti entirely rejects the existence of a 'rate of return' in 'Switches of Technique and "The Rate of Return"', in Capital Theory', *EJ*, 1969, 508–31. See also M. Arcelli, 'La controversia sul capitale e la teoria neoclassica', *L'Industria*, 1970, 299–314.

82   H. S. Houthakker, 'The Pareto Distribution and the Cobb-Douglas Production Function in Activity Analysis', *RES*, 1955, 27–31.

83   R. M. Solow, 'Some Recent Developments in the Theory of Production', in *The Theory and Empirical Analysis of Production*, ed. M. Brown (New York, 1967), pp. 46–8.

84   D. Levhari, 'A Note on Houthakker's Aggregate Production Function in the Multifirm Industry', *Econometrica*, 1968, 151–4.

85   R. Sato, 'Micro- and Macro-Constant Elasticity of Substitution Production Functions in a Multifirm Industry', *JET*, 1969, 438–53.

86   L. Johansen, *Production Functions, An Integration of Micro and Macro, Short Run and Long Run Aspects* (Amsterdam, 1972).

87   C. R. Hulton, 'Technical Change and the Reproducibility of Capital', *AER*, 1975, 956–65.

88   See also G. C. Harcourt, 'The Cambridge Controversies: Old Ways and New Horizons—or Dead End?', *OEP*, 1976, 25–65 and, for another view, C. J. Bliss, *Capital Theory and the Distribution of Income* (Amsterdam, 1975).

CHAPTER 10: TECHNICAL CHANGE AND ECONOMIC GROWTH

1   *The Collected Scientific Papers of Paul A. Samuelson*, vol. 3, ed. R. C. Merton (Cambridge, Mass., 1972), p. 691.

2   Cf. also C. C. von Weizsäcker, *Zur ökonomischen Theorie des technischen Fortschritts* (Göttingen, 1969), pp. 50ff.

3   For another interpretation of embodied technical change, see H. A. Green, 'Embodied Progress, Investment and Growth', *AER*, 1966, 138–51.

4   R. M. Solow, 'Investment and Technical Progress', in *Mathematical Methods in the Social Sciences* (Stanford, 1960), pp. 89–104, reprinted in *Readings in the Modern Theory of Economic Growth*, ed. J. E. Stiglitz and H. Uzawa (Cambridge, Mass., 1969), pp. 156–71. An interesting study that has not received enough attention is B. F. Massell, 'Investment, Innovation and Growth', *Econometrica*, 1962, 239–52. Mention should also be made of A. R. Begstrom, 'A Model of Technical Progress, the

Production Function and Cyclical Growth', *Economica*, 1962, 357–70.

5   Levhari has studied more general production functions in the case of Harrod-neutral technical change embodied in capital goods. D. Levhari, 'Exponential Growth and Golden Rules in Vintage Capital Models', *Macroeconomica*, 1966, 154–66.

6   For an elegant study, see K. J. Stephens, 'A Growth Model with Capital and Labour Vintages', *Metroeconomica*, 1972, 103–18.

7   The consequences of embodied technical change for the construction of the wage-interest frontier are discussed by D. Levhari and E. Sheshinski, 'The Factor Price Frontier with Embodied Technical Progress', *AER*, 1970, 807–13.

8   The possibility of more than one equilibrium wage rate is ignored here in the framework of a growth model. For this see K. Inada, 'Economic Growth and Factor Substitution', *IER*, 1964, 318–27; M. C. Kemp, E. Sheshinski and P. C. Thanh, 'Economic Growth and Factor Substitution', *IER*, 1967, 243–51.

9   This objection is eliminated by Stiglitz in his model by not aggregating heterogeneous capital goods; J. E. Stiglitz, 'Allocation of Heterogeneous Capital Goods in a Two-Sector Economy', *IER*, 1969, 330–73.

10 Cf., e.g., E. S. Phelps, 'The New View of Investment: A Neoclassical Analysis', *QJE*, 1962, 548–67; R. C. Matthews, 'The New View of Investment: Comment', *QJE*, 1964, 164–76; D. Levhari and E. Sheshinski, 'On the Sensitivity of the Level of Output to Savings: Embodiment and Disembodiment', *QJE*, 1967, 524–8; F. M. Fisher, D. Levhari and E. Sheshinski, 'On the Sensitivity of the Level of Output to Savings: Embodiment and Disembodiment, A Clarifactory Note', *QJE*, 1969, 347 and 348; cf. also A. B. Atkinson, 'On Embodiment and Savings', *QJE*, 1970, 127–33. See also H. Borger, 'Embodied Versus Disembodied Improvements', *RE and S*, 1976, 372–5.

11 An important extension of Solow's model has been developed by Westfield, who assumes diminishing and increasing returns to scale: F. M. Westfield, 'Technical Progress and Returns to Scale', *RE and S*, 1966, 432–41. M. D. McCarthy has used Solow's approach in the case of a CES production function: M. D. McCarthy, 'Embodied and Disembodied Technical Progress in the Constant Elasticity of Substitution Production Function', *RE and S*, 1963, 71–5.

12 C.f., e.g., L. Johansen, 'Substitution versus Fixed Production Coefficients in the Theory of Economic Growth: A Synthesis', *Econometrica*, 1959, 157–75; R. M. Solow, 'Substitution and Fixed Proportions in the Theory of Capital', *RES*, 1962, 207–18; E. S. Phelps, 'Substitution, Fixed Proportions, Growth and Distribution', *IER*, 1963, 265–88; C. J. Bliss, 'On Putty-Clay', *RES*, 1968, 105–32; P. K. Bardhan, 'Equilibrium Growth in a Model with Economic Obsolescence of Machines', *QJE*, 1969, 312-23, and P. K. Bardhan, 'More on Putty-Clay', *IER*, 1973, 211–22. Another article of importance is S. C. Hu, 'Putty-Putty versus Putty-Clay: A Synthesis', *IER*, 1972, 324-41.

13 J. R. Hicks, *Theory of Wages*, p. 124.

14 Ibid., p. 125; cf. also J. Robinson, *The Accumulation of Capital* (London, 1956), 169–71.

15 W. E. Salter, op. cit., p. 43. Salter's view is confirmed in an empirical study by J. A. Rasmussen, 'Applications of a Model of Endogenous Technical Change to US Industry Data', *RES*, 1973, 225–38.

16 W. J. Fellner, 'Two Propositions in the Theory of Induced Innovation', *EJ*, 1961, 305–8, reprinted in *The Economics of Technological Change*, ed. N. Rosenberg (Harmondsworth, 1971), pp. 203ff. Cf. also Fellner, 'Measures of Technological Progress in the Light of Recent Growth Theories', *AER*, 1967, 1073–98.

17 C. Kennedy, 'Samuelson on Induced Innovation', *RE and S*, 1966, 442.

18 C. Kennedy, 'Induced Bias in Innovation and the Theory of Distribution', *EJ*, 1964, 541, reprinted in *Readings in the Modern Theory of Economic Growth*, ed. J. E. Stiglitz and H. Uzawa, pp. 149ff.

19 This can also be seen from W. D. Nordhaus, 'Some Skeptical Thoughts on the Theory of Induced Innovations', *QJE*, 1973, 210.

20 Cf., e.g., S. Ahmad, 'On the Theory of Induced Inventions', *EJ*, 1966, 344ff., which is a critical verbal discussion of Kennedy's contribution; see also P. A. Samuelson, 'The Theory of Induced Innovation Along Kennedy-Weizsäcker Lines', *RE and S*, 1965, 343ff., E. M. Drandakis and E. S. Phelps, 'A Model of Induced Inventions, Growth and Distribution', *EJ*, 1966, 823; and C. C. von Wiezsäcker, 'Tentative Notes on a Two-Sector Model with Induced Technical Progress', *RES*, 1966, 245ff.; another article of importance is M. I. Kamien and N. L. Schwartz,, 'Optimal "Induced" Technical Change', *Econometrica*, 1968, 1–17.

21 K. Lancaster, *Consumers Demand* (New York, 1971).

22 R. A. McCain, 'Induced Bias in Technical Innovation Including Product Innovation in a Model of Economic Growth', *EJ*, 1974, 959–66.

23 For this see H. Hollander, 'Eine einfache Begründung zur langfristigen Harrod-Neutralität des technischen Fortschritts,' *Zeitschrift für die gesamte Staatswissenschaft*, 1969, 236–42. Some refinements are discussed in M. Borchert, 'Eine "einfache" Begründung zur langfristigen Harrod-Neutralität des technischen Fortschritts', *Zeitschrift für die gesamte Staatswissenshaft*, 1971, 1–6, and M. Braulke and V. Menschow, 'Induzierter technischer Fortschritt', *Zeitschrift für die gesamte Staatwissenshaft*, 1971, 200–12.

24 W. W. Chang, 'A Model of Economic Growth with Induced Bias in Technical Progress', *RES*, 1972, 205–12. The same author has introduced the share-oriented theory of induced inventions into a neoclassical two-sector model of economic growth: W. W. Chang, 'Induced Bias Invention and Two-Sector Growth Model', *Metroeconomica*, 1973, 288–305.

25 W. D. Nordhaus, 'The Optimal Rate and Direction of Technical Change', in *Essays on the Theory of Optimal Economic Growth*, ed. K. Shell (Cambridge, Mass., 1967), p. 53; *idem, Invention, Growth and*

Welfare, A Theoretical Treatment of Technological Change (Cambridge, Mass., 1960); *idem*, 'An Economic Theory of Technological Change', *AER*, 1969, 18–28.

26 K. Shell, 'A Model of Inventive Activity and Capital Accumulation', in *Essays on the Theory of Optimal Economic Growth*, pp. 67ff. See also R. Eisen, ' "Forschunginduzierter" technischer Fortschritt und Kapitalakkumulation in einem neoclassischen Wachstumsmodell', *Jahrbücher für Nationalökonomie und Statistik*, 1974, 97–118.

27 K. J. Arrow, 'The Economic Implications of Learning by Doing', *RES*, 1962, 155–73; cf. also D. Levhari, 'Further Implications of Learning by Doing', *RES*, 1966, 31–8; *idem*, 'Extensions of Arrow's "Learning by Doing" ', *RES*, 1966, 117–31; E. Sheshinski, 'Optimal Accumulation with Learning by Doing', in *Essays in the Theory of Optimal Economic Growth*, ed. K. Shell, pp. 31–52. On the connection between the learning process and the optimal rate of saving, see also J. Black, 'Learning by Doing and Optimum Savings', *CJEPS*, 1969, 604–12. See also W. J. Fellner, 'Specific Interpretations of Learning by Doing', *JET*, 1969, 119–40.

28 On the significance of the learning process for growth in general, see J. Niedereichholz, 'Die Berücksichtingung von Lernprozessen in Wirtschaftsmodellen', *Jahrbücher für Nationalökonomie und Statistik*, 1970, 114–25. Labour-augmenting technical development in learning models has been called 'Arrow-type' technical development; cf. K. Inada, 'Endogenous Technical Progress and Steady Growth', *RES*, 1969, 99, and also J. S. Lane, 'The Implications of Steady State Growth for Endogenous and Embodied Technical Change', *IER*, 1972, p. 343.

29 Fu-Sen Chen, 'Induced Embodied Technical Progress Possibility Frontier, An Alternative Extension of "Learning by Doing" ' (unpublished manuscript). The same author had previously made a contribution to an endogenous approach to technical development, see Fu-Sen Chen, 'Technical Adventures, Technical Heritage and the Rate of Technical Progress', *Metroeconomica*, 1971, 227–32.

30 For this see W. Y. Oi, 'The Neoclassical Foundations of Progress Functions', *EJ*, 1967, 579–94.

31 T. W. Schultz, Capital Formation on Education, *JPE*, 1960, 571–87; *idem*, 'Investment in Human Capital', *AER*, 1961, 5 and 6; *idem*, *Reflection on Investment in Human Capital, The Role of Education and of Research* (London, 1971).

32 H. Uzawa, 'Optimum Technical Change in An Aggregative Model of Economic Growth', *IER*, 1965, 18–31.

33 E. S. Phelps, *Golden Rules of Economic Growth* (New York, 1967), pp. 158–65; see also R. R. Nelson and E. S. Phelps, 'Investment in Humans, Technological Diffusion and Economic Growth', *AER*, 1966, 69–75; W. Lachmann, 'Produzierbarer technischer Fortschritt und optimales Wirtschaftswachstum', *Zeitschrift für die gesamte Staatswissenschaft*, 1976, 63–78.

34 E. S. Phelps, 'Models of Technical Progress and the Golden Rules of

Research', *RES*, 1966, 133–45 and *Golden Rules of Economic Growth*, pp. 137–57.

35 R. Eisen, '"Forschungsinduzierter" technischen Fortschritt und Kapitalakkumulation', *Jahrbücher für Nationalökonomie und Statistik*, 1974, 97–118.

36 W. A. Eltis, 'The Determination of the Rate of Technical Progress', *EJ*, 1971, 502. See also B. R. Williams, 'Investment and Technology in Growth', *The Manchester School of Economic and Social Studies*, 1964, 59–78.

37 R. E. Lucas, 'Test of a Capital-Theoretic Model of Technological Change', *RES*, 1967, 175. Cragg says that changes in technology are 'investment-complemented' to a greater or lesser extent: J. G. Cragg, 'Technological Progress, Investment and Growth', *CJEPS*, 1963, 319.

38 Equilibrium growth paths can also be derived on the basis of variable returns to scale; see, e.g., J. Conlisk, 'Non-Constant Returns to Scale in a Neoclassical Growth Model', *IER*, 1968, 369–73; *idem*, 'Non-Constant Returns to Scale and the Technical Change Frontier', *QJE*, 1971, pp. 483–93. The existence and stability of equilibrium growth paths in a two-sector model have been examined by W. W. Chang in 'A Two-Sector Model of Economic Growth with Induced Technical Progress', *CJEPS*, 1970, 199–212.

39 For this see e.g. J. Conlisk, 'A Neoclassical Growth Model with Endogeneously Positioned Technical Change Frontier', *EJ*, 1969, 360.

40 See H. Brems, *Labour, Capital and Growth* (London, 1973), chapters 9 and 10. R. Doddy and M. Gort, 'Obsolescence, Embodiment and the Explanation of Productivity Change', *SEJ*, 1974, 553–62.

41 H. Adachi, 'Factor Substitution and Durability of Capital in a Two-Sector Putty-Clay Model', *Econometrica*, 1974, 773–801.

42 The role of saving in this connection has been examined by W. W. Chang, 'The Role of Saving in a Growth Model with Induced Invention', *RE and S*, 1970, 62–7.

43 Equilibrium growth means that capital, labour and production all grow at the same rate and the term does not imply any welfare judgment; the situation in the developing countries is a good illustration of this. See, e.g., J. C. Fei and G. Ranis, 'Innovation, Capital Accumulation and Economic Development', *AER*, 1963, 283–313, and *idem*, 'Innovational Intensity and Factor Bias in the Theory of Growth', *IER*, 1965, 182–98. Also of importance are: R. C. Porter, 'Technological Change with Unlimited Supplies of Labour', *The Manchester School of Economic and Social Studies*, 1968, 69–73; S. Y. Park, 'Surplus Labor, Technical Progress, Growth and Distribution', *IER*, 1969, 22–35. A somewhat more general article on various types of technical change is: F. O. Goddard, 'Harrod-neutral Economic Growth and Hicks-Biased Technological Progress', *SEJ*, 1969/70, 300–8. Another important publication is A. C. Kelley, J. G. Williamson and R. J. Cheetham, 'Biased Technological Progress and Labour Force', *QJE*, 1972, 426–47. A broader interpretation of equilibrium growth is given by Batra, who assumes that the growth rates of labour and capital are constantly

changing, R. N. Batra, 'Technical Progress and Relative Stability of a Two-Sector Model of Economic Growth', *SEJ*, 1971/72, 294.

44 R. M. Solow, J. Tobin, C. C. von Weizsäcker and M. Yaari, 'Neoclassical Growth with Fixed Factor Proportions', *RES*, 1966, 79–115. Inada discusses the opposite question, i.e. whether Harrod-neutral technical change is the only type of exogenous disembodied technical change that is compatible with such a model with an equilibrium growth rate; see K. Inada, 'Fixed Factor Coefficients and Harrod-Neutral Technical Progress', *RES*, 1969, 89–97.

45 R. Sato, 'Fiscal Policy in a Neo-Classical Growth Model: An Analysis of Time required for Equilibrating Adjustment', *RES*, 1963, 16–23; *idem*, 'The Harrod-Domar Models vs. the Neo-Classical Growth Model', *EJ*, 1964, 380–7; J. Conlisk, 'Unemployment in a Neoclassical Growth Model: The Effect on Speed of Adjustment', *EJ*, 1960, 550–6.

46 K. Sato, 'On the the Adjustment Time in Neo-Classical Growth Models', *RES*, 1963, 263–8. Cf. also R. Rawanathan, 'Adjustment Time in the Two-Sector Growth Model with Fixed Coefficients', *EJ*, 1973, 1236–44.

47 E. S. Phelps, 'The New View of Investment: A Neo-Classical Analysis', *QJE*, 1962, 548–67, reprinted in *Readings in the Modern Theory of Economic Growth*, ed. J. E. Stiglitz and H. Uzawa, pp. 172–91.

48 Cf., e.g., R. C. Matthews, 'The New View of Investment: Comment', *QJE*, 1964, 164–72; E. S. Phelps and M. Yaari, 'The New View of Investment: A Reply', *QJE*, 1964, 172–6; D. Levhari and E. Sheshinski, 'On the Sensitivity of the Level of Output to Savings: Embodiment and Disembodiment', *QJE*, 1967, 524–8; F. M. Fisher, D. Levhari and E. Sheshinski, 'On the Sensitivity of the Level of Output to Savings: Embodiment and Disembodiment: A Clarifacatory Note', *QJE*, 1969, 347–8; D. Levhari and E. Sheshinski, 'The Relation between the Rate of Return and the Rate of Technical Progress', *RES*, 1969, 363–79. See also R. M. Lansing, 'On Technical Progress and the Speed of Adjustment', *Economica*, 1975, 394–401.

49 M. Brown, *On the Theory and Measurement of Technological Change* (Cambridge, 1966), p. 59.

50 For this see W. D. Nordhaus, *Invention, Growth and Welfare, A Theoretical Treatment of Technological Change*, pp. 116–39.

51 In this connection, see an important article by J. Kirker Stephens, 'Non-Neutral Technological Progress and Nonhomogeneous Labour in a Cobb-Douglas Growth Model', *Economia Internazionale*, 1971, 446–55.

52 G. Akerlof and W. D. Nordhaus, 'Balanced Growth—A Razor's Edge', *IER*, 1967, 343. Cf. also A. Chilosi and S. Gomulka, 'Progresso tecnico e sviluppo di lungo periodo' ('Technical Progress and Long-Term Development'), *Rivista di politica economica*, 1969, 623–61.

53 In this connection see Rostow on the 'leading sectors', W. W. Rostow, *The Stages of Economic Growth* (London, 1960) and *The Process of Economic Growth*, 2nd ed. (New York, 1962), pp. 261ff.; and *The*

Economics of Take-Off into Sustained Growth, ed. W. W. Rostow (London, 1963), pp. 1–21.

54 M. Kalecki, *Theory of Economic Dynamics* (London, 1954), pp. 157–61.

55 M. Kalecki, *Selected Essays on the Dynamics of the Capitalist Economy, 1933–1970* (Cambridge, 1971).

56 J. Akerman, *Ökonomischer Fortschritt und ökonomische Krisen* (Vienna, 1932).

57 M. Kalecki, 'Political Aspects of Full Employment', *Political Quarterly*, 1943, 322–31, reprinted in *A Critique of Economic Theory*, ed. E. K. Hunt and J. G. Schwartz (London, 1972), pp. 420–30.

58 J. R. Hicks, 'The Permissive Economy', in *Crisis '75* (London, 1975), p. 20.

59 Cf. also A. Asimakopoulos, 'A Kaleckian Theory of Income Distribution'. *CJEPS*, 1975, 313–33.

60 K. R. Popper, *The Logic of Scientific Discovery* (London, 1959) and *Conjectures and Refutations* (London, 1963).

61 For this see, e.g., A. A. Walters, *An Introduction to Econometrics* (London, 1968), pp. 15–20.

62 For this see, e.g., J. S. Cramer, *Empirical Econometrics* (Amsterdam, 1969), pp. 3–7.

63 Cf. also the somewhat similar classification in C. Kennedy and A. P. Thirlwall, 'Technical Progress: A Survey', *EJ*, 1972, 11–72; *idem*, 'Technical Change and the Distribution of Income', *IER*, 1973, 780–4.

64 H. B. Chenery, 'Engineering Production Functions', *QJE*, 1949, 507–31; also *idem*, 'Process and Production Functions from Engineering Data', in *Studies in the Structure of the American Economy*, ed. W. Leontieff (New York, 1953), pp. 297–325. For a penetrating theoretical account see J. Marsden, D. Pingry and W. Whinston, 'Engineering Foundations of Production Functions', *JET*, 1974, 124–40.

65 Cf. W. Z. Hirsch, 'Manufacturing Process Functions', *RE and S*, 1952, 143–55 and 'Firm Progress Ratios', *Econometrica*, 1956, 136–43.

66 On this see especially L. Johansen, *Production Functions, An Integration of Micro and Macro Short Run and Long Run Aspects* (Amsterdam, 1972), p. 195.

67 For a review article, see E. D. Domar, 'On the Measurement of Technological Change', *EJ*, 1961, 709–29. See also W. P. Hogan, 'Productivity Research and Development', *The Economic Record*, 1966, 61–4 and I. B. Kravis, 'A Survey of International Comparisons of Productivity', *EJ*, 1966, 1–45.

68 For this see, e.g., Y. Mundlak and A. Razin, 'Aggregation, Index Numbers and the Measurement of Technical Change', *RE and S*, 1969, 166–75.

69 H. Menderhausen, 'On the Significance of Douglas' Production Function', *Econometrica*, 1938, 143–53.

70 P. H. Douglas, 'Comments on the Cobb-Douglas Production Function', included in *The Theory and Empirical Analysis of Production*, ed. M. Brown (New York, 1967), p. 18.

71 P. H. Douglas, 'Are There Laws of Production?', *AER*, 1948, 1–41 and

'The Cobb-Douglas Production Function Once Again: its History, its Testing and some Empirical Values', *JPE*, 1976, 903–15.

72 D. Durand, 'Some Thoughts on Marginal Productivity with Special Reference to Professor Douglas' Analysis', *JPE*, 1937, 740–58.

73 See F. M. Fisher, 'Aggregate Production Functions and the Explanation of Wages: A Simulation Experiment', *RE and S*, 1971, 305–25. See also S. F. Chu, D. J. Aigner and M. Frankel, 'On the Log-Quadratic Law of Production', *SEJ*, 1970, 32–9.

74 See note 11, Chapter 8.

75 J. Marschak and W. H. Andrews, 'Random Simultaneous Equations and the Theory of Production', *Econometrica*, 1944, 143–205.

76 R. M. Solow, 'Technical Change and the Aggregate Production Functions', *RE and S*, 1957, 312–20, reprinted in *Growth Economics*, ed. A. K. Sen (Harmondsworth, 1970), pp. 401–19.

77 B. F. Massell, 'Capital Formation and Technological Change in US Manufacturing', *RE and S*, 1960, 182–8.

78 C. G. Archibald, *Investment and Technical Change in Greek Manufacturing* (Athens, 1964); especially pp. 40–58.

79 I. W. McLean, 'Growth and Technological Change in Agriculture: Victoria 1870–1910', *The Economic Record*, 1973, 560–74. See also R. A. Powell, 'Growth and Technical Change in Agriculture: A Comment', *The Economic Record*, 1974, 616–19, and I. W. McLean, 'Growth and Technological Change in Agriculture: A Reply', *The Economic Record*, 1974, 620–2.

80 See, e.g., W. P. Hogan, 'Technical Progress and the Production Functions', *RE and S*, 1958, 407–11 and Solow's reply, pp. 411–3; H. S. Levine, 'A Small Problem in the Analysis of Growth', *RE and S*, 1960, pp. 225–8; R. W. Resek, 'Neutrality of Technical Progress', *RE and S*, 1967, 55–63.

81 M. Abramowitz, 'Resource and Output Trends in the United States Since 1870', *AER*, 1956, 5–23.

82 J. W. Kendrick, *Productivity Trends in the United States* (Princeton, N.J., 1961).

83 E. F. Denison, 'United States Economic Growth', *Journal of Business*, 1962, 109–21; *The Residual Factor and Economic Growth* (Paris, 1962), pp. 13–103; *The Sources of Economic Growth in the United States and the Alternatives Before Us* (New York, 1962); *Why Growth Rates Differ* (Brookings Institute, Washington, 1967). Abramowitz has compared Denison's approach with that of Kendrick's in 'Economic Growth in the United States', *AER*, 1962, 762–82.

84 S. Kuznets, *National Income: A Summary of Findings* (New York, 1946); *Economic Change* (London, 1954); *Postwar Economic Growth* (Cambridge, Mass., 1964); *Modern Economic Growth* (New Haven, 1966), *Economic Growth of Nations* (Cambridge, Mass., 1971). See also A. F. Burns, *Production Trends in the United States Since 1870, (New York, 1934).*

85 For Italy, see e.g. N. Cacace and P. Gardin, *Produttività e divario tecnico*

('Productivity and Technical Change') (Milan, 1968), and for Ireland, K. A. Kennedy, *Productivity and Industrial Growth* (Oxford, 1971).

86   M. Brown and J. Popkin, 'A Measure of Technological Change and Returns to Scale', *RE and S, 1962, 402–11*.

87   J. W. Kendrick and R. Sato, 'Factor Prices, Productivity and Economic Growth', *AER*, 1963, 974–1004.

88   M. Brown and J. S. de Cani, 'Technological Change and the Distribution of Income', *IER*, 1963, 289–309.

89   C. E. Ferguson, 'Cross-section Production Functions and the Elasticity of Substitution in American Manufacturing Industry', *RE and S*, 1963, 305–13.

90   C. E. Ferguson, 'Substitution, Technical Progress and Returns to Scale', *AER*, 1963, 296–305. Similar results have been obtained for Switzerland in the period 1951–68 by J. K. Ardenti and J. P. Reichenbach, 'Estimation de la fonction de production C.E.S. pour la Suisse', *Schweizerische Zeitschrift für Volkswirtschaft und Statistik*, 1972, 575–90. A cross-section-type investigation for Spain, covering a large number of industries, was carried out in 1968 by J. B. Donges: 'Returns to Scale and Factor Substitutability in Spanish Industry', *Weltwirtschafliches Archiv*, 1972, 597–608. Woodfield's work on New Zealand is also of importance: A. Woodfield, 'Estimates of Hicks-Neutral Technical Progress, Returns and the Elasticity of Substitution in New Zealand Manufacturing, 1926–1968', *New Zealand Economic Papers*, 1972, 73–92.

91   C. E. Ferguson, 'Time-Series Production Functions and Technological Progress in American Manufacturing Industry', *JPE*, 1965, 135–47.

92   P. A. David and T. van de Klundert, 'Biased Efficiency Growth and Capital-Labor Substitution in the US, 1899–1960', *AER*, 1965, 377–94. Similar investigations were carried out by Kotowitz for Canadian production in the period 1926–61: Y. Kotowitz, 'On the Estimation of a Non-Neutral CES Production Function', *CJEPS*, 1968, 429–39; *idem*, 'Technical Progress, Factor Substitution and Income Distribution in Canadian Manufacturing 1926–39 and 1946–61', *CJEPS*, 1969, 106–14. Binswanger had repeated these studies for the case of many factors of production, H. P. Binswanger, 'The Measurement of Technical Change Biases with Many Factors of Production', *AER*, 1974, 964–76; other important studies are A. Takayama, 'On Biased Technological Progress', *AER*, 1974, 631–9 and T. P. Lianos, 'Factor Augmentation in Greek Manufacturing, 1959–69', *European Economic Review*, 1976, 15–32.

93   See, e.g., M. Nerlove, 'Recent Empirical Studies of the CES and Related Production Functions', *The Theory and Empirical Analysis of Production*, ed., M. Brown, pp. 92ff. Also M. Morishima and M. Saito, 'An Economic Test of Sir John Hicks' Theory of Biased Induced Inventions', in *Value, Capital and Growth, Papers in Honour of Sir John Hicks*, ed. J. N. Wolfe (Edinburgh, 1968), pp. 415–44.

94   N. S. Revankar, 'Capital-Labor Substitution, Technological Change and Economic Growth: The US Experience 1929–1953', *Metro-*

economica, 1971, 154–76. See also C. A. Know Lovell, 'Estimation and Prediction with CES and VES Production Functions', *IER*, 1973, 676–92.

95 A publication along the same lines is M. D. McCarthy, 'Quantity Augmenting Technical Progress and Two-Factor Production Functions', *SEJ*, 1966, 71.

96 E. F. Denison, 'The Unimportance of the Embodied Question', *AER*, 1964, 90–4. See also Yong Keun You, 'Embodied and Disembodied Technical Progress in the United States, 1929–1968', *RE and S*, 1976, 123–7.

97 D. W. Jorgenson, 'The Embodiment Hypothesis', *JPE*, 1966, 1–17.

98 A. Smithies, 'Comments on R. M. Solow, Technical Progress, Capital Formation and Economic Growth', *AER*, 1962, 92.

99 R. M. Solow, *Capital Theory and the Rate of Return* (F. de Vries Lectures, Amsterdam, 1963), p. 43. See also Denison's review: 'Capital Theory and the Rate of Return', *AER*, 1964, 721–5.

100 On the identification question involved here, see M. R. Wickens, 'Estimation of the Vintage Cobb-Douglas Production Function for the United States, 1900–1960', *RE and S*, 1970, 197–93; J. M. Cheng, 'Technical Change Measurement under the Economic Model Including Neutral and Embodied Technology', *SEJ*, 1970/71, 215–17.

101 See note 4, above.

102 See note 98, above.

103 Z. Griliches, 'The Sources of Measured Productivity Growth: United States Agriculture 1940–1960', *JPE*, 1963, 331–46, reprinted in *The Economics of Technological Change*, ed. N. Rosenberg (Harmondsworth, 1971), pp. 382–411.

104 Solow, op. cit., pp. 78–86.

105 R. R. Nelson, 'Aggregate Production Functions and Medium Range Growth Projection', *AER*, 1964, 575–606.

106 B. Gahlen, *Die Überprüfung produktionstheoretischer Hypothesen für Deutschland (1850–1923)* (Tübingen, 1968), p. 263.

107 M. D. Intrilligator, 'Embodied Technical Change and Productivity in the United States 1929–1958', *RE and S*, 1965, 65–70.

108 G. Szakolczai and J. Stahl, 'Increasing or Decreasing Returns to Scale in the Constant Elasticity of Substitution Production Functions', *RE and S*, 1969, 84–90.

109 E. S. Phelps and C. Phelps, 'Factor-Price Frontier Estimation of a "Vintage" Production Model of the Postwar United States Non-farm Business Sector', *RE and S*, 1966, 251–65.

110 D. W. Jorgenson and Z. Griliches, 'The Explanation of Productivity Change', *RES*, 1967, 249–83, reprinted in *Growth Economics*, ed. A. K. Sen, pp. 420–74. See also R. G. Gregory and D. W. James, 'Do New Factories Embody Best Practice Technology?', *EJ*, 1973, 1133–55.

111 E. Sheshinski, 'Test of the "Learning by Doing" Hypothesis', *RE and S*, 1967, 568–78.

112 E. K. Y. Chen, 'The Empirical Relevance of the Endogenous Technical Progress Function', *Kyklos*, 1976, 256–71.

113 See, e.g., M. J. Beckmann and R. Sato, 'Aggregate Production

Functions and Types of Technical Progress: A Statistical Analysis',
AER, 1969, 88–101; R. Sato, 'The Estimation of Biased Technical
Progress and the Production Functions', *IER*, 1970, 179–208; R. Sato
and M. J. Beckmann, 'Shares and Growth under Factor-Augmenting
Technical Change', *IER*, 1970, 387–98; M. J. Beckmann, R. Sato and M.
Schupack, 'Alternative Approaches to the Estimation of Production
Functions, and of Technical Change', *IER*, 1972, 33–52.

114 R. M. Solow, *Growth Theory; An Exposition* (Oxford, 1970),
p. 7

115 For this see also J. Kornai, *Rush Versus Harmonic Growth* (Amsterdam,
1972).

116 See a pioneering contribution by L. J. Mirman, 'The Steady State
Behaviour of a Class of One Sector Growth Models with Uncertain
Technology', *JET*, 1973, 219–42.

117 S. Kuznetz, *Economic Growth of Nations* (Cambridge, 1971), p. 311.
The composition of production can also play a role in the explanation
of the change in productivity; see W. Nordhaus, 'The Recent Productivity Slowdown', *Brookings Papers on Economic Activity*, 1972, 493–
545.

118 H. Reinoud, 'The Evolution of Job Structures in Europe and North
America', in *Manpower Aspects of Automation and Technical Change*
(OECD, Paris, 1966), pp. 111–41.

119 *The Employment Impact of Technological Change*, Studies Prepared for
the National Commission on Technology, Automation and Economic
Progress (Washington, 1966).

120 For this see, e.g., B. L. Scarfe, 'Multi-Sectoral Growth and
Technological Change', *CJEPS*, 1971, 209–313.

121 This subject is discussed in more detail in M. Bronfenbrenner, *Income
Distribution Theory* (Chicago, 1971). This book has been discussed at
length by C. E. Ferguson and E. J. Nell, 'Two Books on the Theory of
Income Distribution: A Review Article', *JEL*, 1972, 437–53. Another
important book on the same subject is W. A. Eltis, *Growth and
Distribution* (London, 1973).

122 See, e.g., M. J. Beckmann, 'Einkommensverteilung und Wachstum bei
nichtneutralem technischem Fortschritt', *Jahrbücher für Nationalökon-
omie und Statistik*, 1965, 80–9; C. E. Ferguson, 'Neoclassical Theory of
Technical Progress and Relative Factor Shares', *SEJ*, 1968, 490–504.
Another important publication is Jae Won Lee, 'Technological Change
and Functional Distribution: With Special Reference to the
Manufacturing Sector', *The Seoul National University Economic
Review*, 1972, 63–80.

CHAPTER 11: TECHNICAL CHANGE AND MONOPOLY POWER

1 J. A. Schumpeter, *Capitalism, Socialism and Democracy*, 3rd edn (New
York, 1950), p. 83.

2 Ibid., pp. 101–6.

3 The publications in this field include: J. S. Bain, *Industrial Organization*
(New York, 1959; 2nd edn, 1968); *idem, Barriers to New Competition*

(Cambridge, Mass., 1956; 3rd edn, 1968); J. M. Clark, *Competition as a Dynamic Process* (Washington, 1961); J. K. Galbraith, *American Capitalism* (Boston, 1956); E. S. Mason, *Economic Concentration and the Monopoly Problem* (Cambridge, Mass., 1959), particularly pp. 91–101; *Industrial Organization and Economic Development*, Collection of essays dedicated to E. S. Mason, ed. J. W. Markham and G. F. Papanek (Boston, 1970); F. M. Scherer, *Industrial Market Structure and Economic Performance* (Chicago, 1970); W. G. Shepherd, *Market Power and Economic Welfare, An Introduction* (New York, 1972). Simple introductions are R. Caves, *American Industry: Structure, Conduct, Performance* (Englewood Cliffs, N.J., 1967) and D. Needham, *Economic Analysis and Industrial Structure* (New York, 1969).

4   J. M. Blair, *Economic Concentration, Structure, Behavior and Public Policy* (New York, 1972), pp. 2–41.

5   A recent illustration of this is to be found in C. Marfels, 'Relevant Market and Concentration: The Case of the US Automobile Industry', *Jahrbücher für Nationalökonomie und Statistik*, 1973, 209–17.

6   An empirical study on this subject is F. M. Scherer, 'The Determinants of Multi-Plant Operation in Six Nations and Twelve Industries', *Kyklos*, 1974, 124–39.

7   For this difference and its significance for monopoly power, see J. M. Blair, op. cit., pp. 102–7 and W. G. Shepherd, op. cit., pp. 120–3.

8   See, e.g., E. T. Penrose, *The Theory of Growth of the Firm* (Oxford, 1966), pp. 104–52 and H. W. de Jong, *Dynamische concentratie-theorie* (Leiden, 1971), pp. 191ff.

9   See, e.g., J. C. Narver, *Conglomerate Mergers and Market Competition* (London, 1967). For the significance of an immediate rise in profits as a reason for setting up a conglomerate, see W. J. Mead, 'Instantaneous Merger Profit as a Conglomerate Merger Motive', *Western Economic Journal*, 1969, 295–306. For other explanations, see M. Gort, 'An Economic Disturbance Theory of Merger', *QJE*, 1969, 624–42 and D. C. Mueller, 'A Theory of Conglomerate Mergers', *QJE*, 1969, 643–59; D. E. Logue and P. A. Naert, 'A Theory of Conglomerate Mergers, Comment and Extension', *QJE*, 1970, 662–7 (see also D. R. Kamerschen's comments and Mueller's reply).

10   See, on these lines, H. B. Thorelli, 'The Political Economy of the Firm: Basis for a New Theory of Competition', *Schweizerische Zeitschrift für Volkswirtschaft und Statistik*, 1965, 252–3.

11   See, e.g., *The Multinational Enterprise*, ed. J. H. Dunning (London, 1971); also of importance are E. T. Penrose, *The Large International Firm in Developing Countries* (London, 1968) and J. H. Dunning, 'The Determinants of International Production', *OEP*, 1973, 289–336. See also F. H. Fleck and R. Mah Fouz, 'The Multinational Corporation', *Schweizerische Zeitschrift für Volkswirtschaft und Statistik*, 1974, 145–60.

12   For a different view see M. C. Sawyer, 'Concentration in British Manufacturing Industry', *OEP*, 1971, 352–75; see also W. G. Shephard, 'British Industrial Concentration: A Comment', *OEP*, 1972,

432–7, together with Sawyer's reply. Cf. further P. E. Hart, M. A. Utton and G. Walshe, *Mergers and Concentration in British Industry* (Cambridge, 1973); M. A. Utton, 'On Measuring the Effects of Industrial Mergers', *Scottish Journal of Political Economy*, 1974, 13–28; K. D. George and A. Silberston, 'The Causes and Effects of Mergers', *Scottish Journal of Political Economy*, 1975, 179–94.

13 For the concept of workable competition, see J. M. Clark, 'Towards a Concept of Workable Competition', *AER*, 1940, 241–56; also H. Sosnick, 'A Critique of Concepts of Workable Competition', *QJE*, 1958, 380–423.

14 W. D. Nordhaus, *Invention, Growth and Welfare, A Theoretical Treatment of Technological Change*, pp. 36–8.

15 J. E. Meade ignores these in his survey, *The Growing Economy* (London, 1968), pp. 111–17.

16 A. Wolfelsperger's *Les biens collectifs* (Paris, 1969) can be recommended for the theory of collective goods; see also M. H. Peston, *Public Goods and the Public Sector* (London, 1972). A brief summary is given in W. J. Baumol, *Welfare Economics and the Theory of the State*, 2nd edn (London, 1965), pp. 20–2.

17 M. I. Kamien and N. L. Schwartz, 'Induced Factor Augmenting Technical Progress from a Micro-Economic Viewpoint', *Econometrica*, 1969, 668–84, and 'Theory of the Firm with Induced Technical Change', *Metroeconomica*, 1971, 233–56. See also A. P. Jacquemin and J. Thisse, 'Strategy of the Firm and Market Structure: An Application of Optimal Control Theory', in *Market Structure and Corporate Behaviour*, ed. K. Cowling (London, 1972), pp. 63–84.

18 S. Koizumi, 'Technical Progress and Investment', *IER*, 1969, 68–81.

19 H. P. Binswanger, 'A Micro-Economic Approach to Induced Innovation', *EJ*, 1974, 940–58.

20 E. Mansfield, 'Industrial Research and Development Expenditure, Determinants, Prospects and Relation to Size of Firm and Inventive Output', *JPE*, 1964, 319–40; 'Rates of Return from Industrial Research and Development', *AER*, 1969, 310–22; *Industrial Research and Technological Innovation: An Econometric Analysis* (New York, 1968); 'Industrial Research, Research and Development: Characteristics, Costs and Diffusion of Results', *AER*, 1969, 65–79; *Research and Innovation in the Modern Corporation*, pp. 18–46.

21 Cf. also D. Hamberg, 'Invention in the Industrial Research Laboratory', *JPE*, 1963, 95–115.

22 E. Mansfield, *Research and Innovation in the Modern Corporation*, p. 217.

23 See, e.g., F. M. Scherer, 'Corporate Inventive Output, Profits and Growth', *JPE*, 1965, 290–7.

24 W. S. Comanor and F. M. Scherer, 'Patent Statistics as a Measure of Technical Change', *JPE*, 1969, 398.

25 See, e.g., J. R. Minisian, 'Research and Development, Production Functions and Rates of Return', *AER*, 1969, 80–5.

26 See C. Freeman, *The Economics of Industrial Innovation* (London, 1974).

27 For this, see S. K. Nath, *A Reappraisal of Welfare Economics* (London, 1969), pp. 18–9.

28 E. Mansfield, *Microeconomics, A Theory and Applications* (New York, 1970), p. 461.

29 T. Scitovsky, *Welfare and Competition*, revised edn (London, 1971), pp. 229–39, 405–7 and 475.

30 R. M. Cyert and K. D. George, 'Competition, Growth and Efficiency', *EJ*, 1969, 26.

31 H. Leibenstein, 'Allocative Efficiency vs. "X-Efficiency" ', *AER*, 1966, 392–415.

32 Ibid., p. 413; for some more recent refinements, see K. J. Blois, 'Some Comments on the Theory of Inert Areas and the Definition of X-Efficiency', *QJE*, 1974, 681-6 and D. A. Peel, 'A Note on X-Efficiency', *QJE*, 1974, 687–8, together with Leibenstein's reply, 'Comment on Inert Areas and the Definition of X-Efficiency', *QJE*, 1974, 689–91.

33 H. Leibenstein, 'Organizational or Frictional Equilibria, X-Efficiency, and the Rate of Innovation', *QJE*, 1969, 600.

34 Cf. also K. J. Blois, 'A Note on X-Efficiency and Profit Maximization', *QJE*, 1972, 310–12 and 'Comment on the Theory and Measurement of Dynamic X-Efficiency', *QJE*, 1972, 313–26, together with Leibenstein's reply in the same Journal, 327–31.

35 On the connection between X-Efficiency and the diffusion of technical knowledge, see also T. Y. Shen, 'Technology Diffusion, Substitution and X-Efficiency', *Econometrica*, 1973, 263–84.

36 R. A. McCain, 'Competition, Information, Redundancy: X-Efficiency and the Cybernetics of the Firm', *Kyklos*, 1975, 286–308.

37 W. S. Comanor and H. Leibenstein, 'Allocative Efficiency, X-Efficiency and the Measurement of Welfare Losses', *Economica*, 1969, 304–9. See also D. Schwartzman, 'Competition and Efficiency: Comment', *JPE*, 1973, 756–64 and H. Leibenstein, 'Competition and X-Efficiency: Reply', *JPE*, 1973, 765–77.

38 M. A. Crew and C. K. Rowley, 'On Allocative Efficiency, X—Efficiency and the Measurement of Welfare Loss', *Economica*, 1971, 199–203.

39 See, e.g., O. E. Williamson, 'A Dynamic Stochastic Theory of Managerial Behaviour in Prices', in *Issues in Theory, Practice and Public Policy*, ed. A. Phillips and O. E. Williamson (Philadelphia, 1967), p. 230. See also by the same author, 'Managerial Discretion Organization Form and the Multidivision Hypothesis', in *The Corporate Economy*, ed. R. Morris and A. Wood (London, 1971), pp. 354–7. See also S. Lofthouse, 'Recent Literature Relating to the X-Efficiency and Market Structure Relationship', *Zeitschrift für Nationalökonomie*, 1974, 409–23

40 R. Parish and Yew-Kwang Ng, 'Monopoly, X-Efficiency and the Measurement of Welfare Loss', *Economica*, 1972, 304.

41 For an application of the theory to the firm, see M. A. Crew, M. W. Jones-Lee and C. K. Rowley, 'X-Theory Versus Management Discretion Theory', *SEJ*, 1971, 173–84. For the significance of X-

Efficiency for the policy on competition, see M. A. Crew and C. K. Rowley, 'Anti-Trust Policy: Economics versus Management Science', in *Readings in Industrial Economics*, vol. 2, ed. C. K. Rowley (London, 1972), pp. 136–49.

42   See, for a more or less similar conclusion, G. J. Stigler, 'The Xistance of X-Efficiency', *AER*, 1976, 213–16.

43   For these problems, see particularly A. K. Cairncross, K. D. George and A. Silberston in a special issue of *The Economic Journal*, dedicated to E. A. G. Robinson, 1972, 311–530. See also S. J. Prais, 'A New Look at the Growth of Industrial Concentration', *OEP*, 1974, 273–88; W. F. Mueller and L. G. Hamm, 'Trends in Industrial Market Concentration, 1947 to 1970', *RE and S*, 1974, 511–20.

44   See L. G. Goldberg's empirical study, 'Conglomerate Mergers and Concentration Ratios', *RE and S*, 1974, 303–9.

45   A. A. Young, 'Increasing Returns and Economic Progress', *EJ*, 1929, 527–42, reprinted in *Readings in Welfare Economics*, ed. K. J. Arrow and T. Scitovsky (London, 1969), pp. 228–41. This article has recently been discussed in detail by N. Kaldor, 'The Irrelevance of Equilibrium Economics', *EJ*, 1972, 1237–55.

46   M. Merhav, *Technological Dependence, Monopoly and Growth* (Oxford, 1969), p. 87.

47   F. H. Hahn, *On the notions of equilibrium in economics* (Cambridge, 1973).

48   R. R. Nelson and S. G. Winter, 'Neoclassical Evolutionary Theory of Economic Growth: Critique and Prospective', *EJ*, 1974, 903. See also N. Georgescu-Roegen, 'Dynamic Equilibrium and Economic Growth', *Economie Appliquée*, 1974, 529–64; R. R. Nelson and S. G. Winter, 'Growth Theory from an Evolutionary Perspective, The Differential Productivity Puzzle', *AER*, 1975, 338–44.

49   G. Viaciago, 'Increasing Returns and Growth in Advanced Economies, A Re-Evaluation', *OEP*, 1975, 232–40.

50   J. E. La Tourette, 'Economies of Scale and Capital Utilization in Canadian Manufacturing, 1926–1972', *CJEPS*, 1975, 448–55.

51   N. Kaldor, *Causes of the Slow Rate of Economic Growth of the United Kingdom* (Cambridge, 1966); N. Kaldor, 'What is Wrong with Economic Theory?', *QJE*, 1975, 347–57.

52   See, for the management of technical change in general, R. O. Burns, *Innovation: The Management Connection* (London, 1975).

53   On diffusion in general, see S. B. Saul, 'The Nature and Diffusion of Technology', in *Economic Development in the Long Run*, ed. A. J. Youngson (London, 1972), pp. 36–61; for the connection between the multinational company and the diffusion of technical knowledge, an important contribution is K. Pavitt, 'The Multinational Enterprise and the Transfer of Technology', in *The Multinational Enterprise*, ed. J. H. Dunning (London, 1971), pp. 61–83. An empirical study had been done by R. E. Caves, 'Multinational Firms, Competition and Productivity in Host-Country Markets', *Economica*, 1974, 176–93. See also R. H. Mason, 'Some Observations on the Choice of Technology by

Multinational Firms in Developing Countries', *RE and S*, 1973, 349–55. For many interesting details, see *The Diffusion of New Industrial Processes, An International Study*, ed. L. Nabseth and G. F. Ray (London, 1974). See also S. Globerman, 'Technological Diffusion in the Canadian Tool and Die Industry', *RE and S*, 1975, 428–34.

54 See A. Phillips, *Technology and Market Structure* (Lexington, Mass., 1971). Phillips explains changes in market structure mainly by the exogenous development of general knowledge.

55 On this subject, see especially J. K. Galbraith, *American Capitalism* (Boston, 1956), pp. 86–7.

56 R. R. Nelson, 'The Simple Economics of Basic Scientific Research', *JPE*, 1959, 297–306.

57 W. S. Comanor, 'Research and Technical Change in the Pharmaceutical Industry', *RE and S*, 1965, 182–91. Comanor's approach has been improved by J. M. Vernon and P. Gusen, 'Technical Change and Firm Size: The Pharmaceutical Industry', *RE and S*, 1974, 274–302.

58 H. G. Grabowski, 'The Determinants of Industrial Research and Developments: A Study of the Chemical, Drug and Petroleum Industries', *JPE*, 1968, 292–305.

59 A. Wood, 'Diversification, Merger and Research Expenditures: A Review of Empirical Studies', in *The Corporate Economy: Growth, Competition and Innovative Potential*, ed. R. Marris and A. Wood (London, 1971), pp. 428ff.

60 D. Hamberg, 'Size of Firms, Oligopoly and Research: The Evidence', *CJEPS*, 1964, 62–75.

61 D. Hamberg, *Research and Development: Essays on the Economics of Research and Developments* (New York, 1966); see also J. D. Howe and D. G. McFetridge, 'The Determinants of R & D Expenditures', *CJEPS*, 1976, 57–71.

62 D. J. Smyth, J. M. Samuels and J. Tzoannos, 'Patents, Profitability, Liquidity and Firm Size', *Applied Economics*, 1972, 85. See also C. Layton, *Ten Innovations* (London, 1972).

63 E. Mansfield, 'The Speed of Response of Firms to New Techniques', *QJE*, 1963, 290–311.

64 E. Mansfield, *Research and Innovation in the Modern Corporation* (London, 1971), p. 13; see also by the same author, 'Size of Firms, Market Structure and Innovation', *JPE*, 1963, 556–76. See further, O. E. Williamson, 'Innovation and Market Structure'. *JPE*, 1965, 67–73, which deals particularly with the effect of monopoly power.

65 W. Adams, and J. B. Dirlam, 'Big Steel, Invention and Innovation', *QJE*, 1966, 167–89; cf. also J. S. Metcalf, 'Diffusion of Innovation in the Lancashire Textile Industry', *The Manchester School of Economic and Social Studies*, 1970, 145–62.

66 A. K. McAdams, 'Big Steel, Invention and Innovation Reconsidered', *QJE*, 1967, 457–74.

67 W. Adams and J. B. Dirlam, 'Big Steel, Inventions and Innovation: Reply', *QJE*, 1967, 481. See also H. G. Baumann, 'The Determinants of

the Diffusion of Technology and Competitiveness in International Trade: A Comparative Case Study of the US and Canadian Steel Industries', *Economia Internazionale,* 1974, 141–58. See also T. Hastings, 'The Characteristics of Early Adopters of New Technology: an Australian Study', *The Economic Record,* 1976, 239–50.

68 B. Branch, 'Research and Development Activity and Profitability. Distributed Lag Analysis', *JPE,* 1974, 999–1012.

69 M. Teubal, 'Threshold R & D levels in Sectors of Advanced Technology', *European Economic Review,* 1976, 395–402.

70 Cf. also Shepherd's conclusion that 'the burden of the proof has now shifted against the larger corporations'. *Market Power and Economic Welfare* (New York, 1972), p. 173.

71 The following publications on the theory of oligopoly recommended: J. S. Bain, *Barriers to New Competition* (Cambridge, Mass., 1956); W. J. Baumol, *Business Behavior, Value and Growth,* 3rd edn (New York, 1967); W. J. Fellner, *Competition Among the Few* (New York, 1949); F. Machlup, *The Economics of Sellers' Competition* (Baltimore, 1952); M. Nicholson, *Oligopoly and Conflict. A Dymamic Approach* (Liverpool, 1972); S. S. Sengupta, *Operations Research in Seller's Competition* (New York, 1967); R. Sherman, *Oligopoly, An Empirical Approach* (Toronto, 1972); L. G. Telser, *Competition, Collusion and Game Theory* (London, 1971).

72 E. H. Chamberlin, *The Theory of Monopolistic Competition,* 6th edn (London, 1948), pp. 46ff. On the connection between Chamberlin's and Fellner's theories, see also J. S. Bain, 'Chamberlin's Impact on Microeconomic Theory', in *Monopolistic Competition Theory: Studies in Impact, Essays in Honor of Edward H. Chamberlin,* ed. R. E. Kuenne (New York, 1967), pp. 165–7.

73 See, e.g., M. Shubik, 'A Further Comparison of Some Models of Duopoly', *Western Economic Journal,* 1968, 264ff.

74 Cf. J. W. Friedman, 'A Noncooperative View of Oligopoly', *IER,* 1971, 106–22. A dynamic study in which the setting-up of stocks of goods is taken into account is A. P. Kirman and M. J. Sobel, 'Dynamic Oligopoly with Inventories', *Econometrica,* 1974, 279–88.

75 O. E. Williamson, 'A Dynamic Theory of Interfirm Behavior', *QJE,* 1965, 579–607. See also J. F. Franke, 'Advertising and Collusion in Oligopoly', *Jahrbücher für Nationalökonomie und Statistik,* 1973, 33–50; R. E. Keunne, 'Towards an Operational General Equilibrium Theory with Oligopoly, some Experimental Results and Conjectures', *Kyklos,* 1974, 792–820.

76 R. B. Heflebower, 'The Theory and Effects of Nonprice Competition', in *Monopolistic Competition Theory,* ed. R. E. Keunne, p. 201.

77 P. Kotler, *Marketing Decision-Making: A Model-Building Approach* (New York, 1971), p. 227.

78 See D. Needham, *Economic Analysis and Industrial Structure* (New York, 1969), p. 95.

79 Pioneering works in this field are: J. S. Bain, *Barriers to New Competition*

and P. Sylos-Labini, *Oligopoly and Technical Progress* (Cambridge, Mass., 1969).

80 See O. E. Williamson, 'Selling Expense as a Barrier to Entry', *QJE*, 1963, 112–28.

81 For the differences that can then be observed in commericial policy, see also R. F. Lanzilotti, *Pricing, Production and Marketing Policies of Small Manufacturers* (Washington, 1964).

82 P. Hennipman, 'Monopoly: Impediment or Stimulus to Economic Progress?', in *Monopoly and Competition and their Regulation*, ed. E. H. Chamberlin (London, 1954), p. 454.

83 For these arguments, see Hennipman's article mentioned above and G. C. Allen, *Monopoly and Restrictive Practices* (London, 1968), p. 23; van der Burg, 'Research en ontwikkeling' ('Research and Development'), VII, *Economisch-Statistische Berichten*, 1969, 724–8; R. Caves, *American Industry: Structure, Conduct, Performance* (Englewood Cliffs, New Jersey, 1967), pp. 100ff.; D. Needham, *Economic Analysis and Industrial Structure* (New York, 1969), pp. 164ff.; F. M. Scherer, *Industrial Market Structure and Economic Performance* (Chicago, 1970), pp. 346ff.; H. Townsend, *Scale, Innovations, Merger and Monopoly* (Oxford, 1968), pp. 25ff.; D. A. Worcester Jr., *Monopoly, Big Business and Welfare in the Post-War United States* (Seattle and London, 1967), pp. 156ff.

84 G. C. Means, 'Administered Prices' in *Economics in Action*, ed. S. M. Mark and D. M. Slate, (San Francisco, 1961), p. 31. See also A. Heertje, 'Prijstheorie en prijspolitiek van de overleid' ('Price Theory and the Price Policy of the Government'), *De Economist*, 1962, 325ff.

85 Cf. also the conclusions of W. Sellekaerts and R. Lesage, 'A Reformulation and Empirical Verification of the Administered Prices Inflation Hypothesis: The Canadian Case', *SEJ*, 1973, 345–60.

86 See also A. S. Eichner, 'A Theory of the Determination of the Mark-Up under Oligopoly', *EJ*, 1973, 1184–1200.

87 For the connection between commercial policy and technical change, see also S. Lombardini, 'Modern Monopolies in Economic Development', in *The Corporate Economy*, ed. R. Marris and A. Wood, pp. 263–6.

88 M. I. Kamien and N. L. Schwartz, 'On the Degree of Rivalry for Maximum Innovative Activity', *QJE*, 1976, 245–60. See also F. M. Scherer, 'Research and Development Resource Allocation under Rivalry', *QJE*, 1967, 359–94; L. E. Ruff, 'Research and Technological Progress in a Cournot Economy', *JET*, 1968, 397–415.

89 Cf. also T. Burns and G. M. Stalker, *The Management of Innovation* (London, 1966), pp. 210ff.

90 For this, see Y. Barzel, 'Optimal Timing of Innovations', *RE and S*, 1968, 348–55.

91 M. I. Kamien and N. L. Schwartz, 'Timing of Innovations under Rivalry', *Econometrica*, 1972, 43–60; cf. also Salter, *Productivity and Technical Change*, pp. 90–92.

92 See also P. L. Swan, 'Market Structure and Technological Progress: The Influence of Monopoly on Product Innovation', *QJE*, 1970, 627–38 and

L. J. White, 'A Note on the Influence of Monopoly on Product Innovation', *QJE*, 1972, 342–9. On the influence of the market structure on excess capacity, see F. Ferguson Esposito and L. Esposito, 'Excess Capacity and Market Structure', *RES*, 1974, 188–94.

93 I. Horowitz, 'Firm Size and Research Activity', *SEJ*, 1962, 298–301.

94 F. M. Scherer, 'Firm Size, Market Structure, Opportunity and the Output of Patented Inventions', *AER*, 1965, 1097–125; *idem*, 'Market Structure and the Employment of Scientists and Engineers', *AER*, 1967, 524–31; 'Research and Development Resource Allocation Under Rivalry', *QJE*, 1967, 385–9.

95 For this see also F. M. Fisher and P. Temin, 'Return to Scale in Research and Development: What Does the Schumpeterian Hypothesis Imply?', *JPE*, 1973, 56–70.

96 A Recent study has been done by R. McGuckin: 'Entry, Concentration Change and Stability of Market Shares', *SEJ*, 1971/72, 363–70.

97 B. Carlsson, 'The Measurement of Efficiency in Production: An Application to Swedish Manufacturing Industries, 1968', *The Swedish Journal of Economics*, 1972, 468–85. See also F. R. Førsund and L. Hjalrmarsson, 'On the Measurement of Productive Efficiency', *The Swedish Journal of Economics*, 1974, 141–54; J. Richmond, 'Estimating the Efficiency of Production', *IER*, 1974, 515–21.

98 J. E. Tilton, *International Diffusion of Technology: The Case of Semi-Conductors* (Washington, 1971), p. 170.

99 Cf., e.g., Barzel's study on the average speed in motor racing, which he uses as a measure of technical development. Y. Barzel, 'The Rate of Technical Progress: The "Indianapolis 500" ', *JET*, 1972, 81–2.

100 These case studies are exemplified by R. L. Samsom, 'The Motor Pump: A Case Study of Innovation and Development', *OEP*, 1969, 109–21.

101 For this see, e.g., I. M. M. Grossack, 'The Concept and Measurement of Permanent Industrial Concentration', *JPE*, 1972, 745–60; C. Marfels, 'On Testing Concentration Measures', *Zeitschrift für Nationalökonomie*, 1972, 461–86; *idem*, 'The Consistency of Concentration Measures: A Mathematical Evaluation', *Zeitschrift für die gesamte Staatswissenschaft*, 1972, 196–215.

102 See, e.g., P. J. Dhrymes and M. Kurz, 'Technology and Scale in Electricity Generation', *Econometrica*, 1964, 287–315. See also M. Galatin, *Economics of Scale and Technological Change in Thermal Power Generation* (Amsterdam, 1968); L. J. Lau and S. Tamura, 'Economics of Scale, Technical Progress and the Nonhomothetic Leontief Production Functions: An Application to the Japanese Petrochemical Processing Industry', *JPE*, 1972, 1167–87.

103 See also M. I. Kamien and N. L. Schwartz, 'Market Structure and Innovation: A Survey', *JEL*, 1975, 1–37.

104 An interesting theoretical analysis of such redeployment can be found in Y. Weiss, 'Learning by Doing and Occupational Specialization', *JET*, 1971, 189–98. A great deal of concrete information has been given by

L. C. Hunter, G. L. Reid and D. Boddy, *Labour Problems of Technological Change* (London, 1970).

105 Some important relevant data are to be found in *Trade Unions and Technological Change*, ed. S. D. Anderman (London, 1966), particularly pp. 138–57.

106 On this topic see especially D. Davies and O. McCarthy, *Introduction to Technological Economics* (London, 1967), particularly pp. 93–109.

107 R. M. Cyert and K. D. George, 'Competition, Growth and Efficiency', *EJ*, 1969, 23–41. See also *Technological Change and Management: The John Diebold Lectures, 1968–1970*, ed. D. W. Ewing (Boston, 1970).

108 P. Sylos-Labini, *Trade Unions, Inflation and Productivity* (Westmead, 1974), p. 78.

109 See also R. R. Nelson and S. G. Winter, 'Toward an Evolutionary Theory of Economic Capabilities', *AER*, 1973, 440–9. The same conclusion is reached by D. S. Johnson in his recent book *The Economics of Invention and Innovation* (London, 1975).

CHAPTER 12: TECHNICAL DEVELOPMENT AND ECONOMIC POLICY

1 P. Hennipman, 'Doeleinden en criteria der economische politiek' ('Objectives and Criteria of Economic Policy'), in *Theorie van de economische politiek* (Leiden, 1962), p. 7.

2 See L. M. Gatovsky, 'Innovation et profit en URSS', *Economie Appliquée*, 1972, 749–814; also T. S. Khachaturov, 'Development of Science and Technology in the USSR', in *Science and Technology in Economic Growth*, ed. B. R. Williams (London, 1973), pp. 147–68; R. N. Batra, 'Technological Change in the Soviet Collective Farm', *AER*, 1974, 594–603.

3 E. Asher and T. K. Kumar, 'Capital-Labor Substitution and Technical Progress in Planned and Market Oriented Economics: A Comparative Study', *SEJ*, 1973, 103–9.

4 See E. A. Hewett, 'The Economics of East European Technology Imports from the West', *AER*, 1975, 377–82.

5 For the measurement of this, see C. Seidl, 'On Measurement of Convergence of Economic Systems', *Zeitschrift für Nationalökonomie*, 1969, 427–32.

6 J. Tinbergen, 'The Theory of the Optimum Regime', in *Selected Economic Papers* (Amsterdam, 1959), pp. 264–304.

7 For an interesting comparative study of growth in the two economic systems, see E. Asher and T. K. Kumar, op. cit.

8 Cf. E. F. Hekscher, *Mercantilism*, vol. 2 (London, 1935), pp. 141ff.

9 Important publications on the economic theory of patents include: A. Plant, 'The Economic Theory Concerning Patents for Inventions', *Economica*, 1934, 30–51; F. Machlup, *Patentwesen, Handwörterbuch der Sozialwissenschaften*, vol. 8 (Tübingen, 1964), pp. 231–52; F. M. Scherer, *Industrial Market Structure and Economic Performance* (Chicago, 1970), pp. 379–99; a good summary has been given by R. Caves, *American Industry: Structure, Conduct, Performance*, pp. 89–92.

It is also well worth reading M. W. Balz, *Invention and Innovation under Soviet Law* (London, 1975). See also H. G. Johnson, 'Aspects of Patents and Licenses as Stimuli to Innovation', *Weltwirtschaftliches Archiv*, 1976, 417–28.

10 F. Machlup, *The Economics of Sellers' Competition* (Baltimore, 1952), p. 556.

11 F. List, *Das nationale System der politischen Oekonomie* (Stuttgart and Tübingen, 1841), pp. 427–36. See also C. F. Bastable, *The Theory of International Trade*, 4th edn (London, 1929). A good summary is to be found in P. B. Kenen, *International Economics* (Englewood Cliffs, New Jersey, 1964). Also recommended is H. Randak, *Freidrich List und die wissenschaftliche Wirtschaftpolitik*, (Tübingen, 1972), pp. 39–49.

12 Cf., e.g., H. G. Grubel, 'The Anatomy of Classical and Modern Infant Industry Arguments', *Weltwirtschafliches Archiv*, 1966, vol. 2, 325–44; R. E. Baldwin, 'The Case against Infant-Industry Tariff Protection', *JPE*, 1969, 295–305.

13 See M. V. Posner, 'International Trade and Technical Change', *OEP*, 1961, 323–30; G. C. Hufbauer, *Synthetic Materials and the Theory of International Trade (London, 1966); R. Vernon, 'International Investment and International Trade in the Product Cycle', QJE*, 1966, 191–207; S. Hirsch, *Location of Industry and International Competitiveness* (London, 1967).

14 See F. Machlup, 'The Optimum Lag of Imitation behind Innovation', in *Festschrift für Frederik Zeuthen*, ed. P. Milhøj (Copenhagen, 1958).

15 J. N. Bhagwati, 'The Pure Theory of International Trade: A Survey', *EJ*, 1964, 1–84, reprinted in *Surveys of Economic Theory*, vol. 2 (London, 1968).

16 J. S. Chipman, 'A Survey of the Theory of International Trade', *Econometrica*, 1965, 477–519 and 685–760.

17 See, e.g., his *Money, Trade and Growth* (London, 1962), pp. 75ff. and a recent contribution in the same vein, 'Trade and Growth: A Geometrical Exposition', *Journal of International Economics*, 1971, 83–101 and 'Trade and Growth: A Correction', *Journal of International Economics*, 1972, 87–8. This article has also been included in a volume dedicated to C. P. Kindleberger, entitled *Trade, Balance of Payments and Growth* (Amsterdam, 1971), pp. 144–67; see also his recent book *Technology and Economic Interdependence* (London, 1975).

18 I. F. Pearce, *International Trade* (London, 1970), especially chapter 10, vol. 1 (pp. 268–80) and chapter 18, vol. 2 (pp. 568–615); M. C. Kemp, *The Pure Theory of International Trade*, 2nd edn (London, 1969), pp. 111–13 and 282–5; *idem*, 'Technological Change, The Terms of Trade and Welfare', *EJ*, 1955, 457–73, which can be described as a pioneering study. Also of interest: B. Södersten, *International Economics* (London, 1970), pp. 160–83; S. B. Linder, 'Trade and Technical Efficiency', in *Trade, Balance of Payments and Growth*, pp. 495–506; N. C. Miller and N. L. Schwartz, 'Factor-Augmenting Technical Advance in a Two-Sector Open Economy', *OEP*, 1970, 338–56; D. D. Purvis, 'Technology, Trade and Factor Mobility', *EJ*, 1972, 991–9; R. W. Klein, 'A Dynamic

Theory of Comparative Advantage', *AER*, 1973, 173–84: Wontack Hong, 'A Dynamic Model of International Trade with Endogenous Technical Progress', *The Seoul National University Economic Review*, 1972, 31–50; M. Teubal, 'Comparative Advantage and Technological Change: The Learning by Doing Case', *Journal of International Economics*, 1973, 161–77; Teubal has further elaborated this topic in 'A Neotechnology Theory of Cpmparative Costs', *QJE*, 1975, 414–31; G. W. Scully, 'Technical Progress, Factor Market Distortions and the Pattern of Trade', *Economia internazionale*, 1973, 3–16; S. Hirsch, 'Capital or Technology? Confronting the Neo-Factor Proportions and Neo-Technology Accounts of International Trade', *Weltwirtschaftliches Archiv*, 1974, 535–63; L. Dudley, 'Learning and the Interregional Transfer of Technology', *SEJ*, 1969, pp. 563–70; J. R. Melvin, 'Technological Change, Factor Intensity Reversals and Trade,' *Economica*, 1976, 173–80.

19   Cf., e.g., W. Gruber, D. Mehta and R. Vernon, 'The R and D Factor in International Trade and International Investment of United States Industries', *JPE*, 1967, 20–37; see also D. B. Keesing, 'The Impact of Research and Development on United States Trade', *JPE*, 1962, 38–48; J. R. Artus, 'The Short-Run Effects of Domestic Demand Pressure on Export Delivery Delays for Machinery', *Journal of International Economics*, 1973, 21–36. An important empirical study has been published by H. Katrak, 'Human Skills, R and D and Scale Economies in the Export of the United Kingdom and the United States', *OEP*, 1973, 337–60. T. C. Lowinger, 'The Technology Factor and the Export Performance of US Manufacturing Industry', *Economic Inquiry*, 1975, 221–36.

20   C. A. Rodriguez, 'Trade in Technological Knowledge and the National Advantage', *JPE*, 1975, 121–35. See also J. Borkakoti, 'Some Welfare Implications of the Neo-technology Hypothesis of the Pattern of International Trade', *OEP*, 1975, 383–400.

21   For the significance of the granting of patents in this connection, see E. Penrose, 'International Patenting and the Less Developed Countries', *EJ*, 1973, 768–85.

22   J. Tinbergen, *Shaping the World Economy* (New York, 1962), pp. 119ff. Another important work is P. Streeten, 'Technological Gaps between Rich and Poor Countries', *Scottish Journal of Political Economy*, 1972, 213–30.

23   S. Chakravarty, *Capital and Development Planning* (Cambridge, Mass., 1969).

24   H. J. Bruton, *Principles of Development Economics* (Englewood Cliffs, New Jersey, 1965), pp. 348ff.; cf. also H. Leibenstein, *Economic Backwardness and Economic Growth* (New York, 1957), pp. 254ff.

25   Employment in the short and long term has been discussed by J. Encarnación, 'On appropriate Technology, Saving and Employment', *Journal of Development Economics*, 1974, 71–9. Cf. also V. E. Tokman, 'Income Distribution, Technology and Employment in Developing

Countries: An Application to Ecuador', *Journal of Development Economics*, 1974, 49–80.

26 Attempts at quantifying are illustrated by G. S. Maddala and P. T. Knight, 'International Diffusion of Technical Change, A Case Study of the Oxygen Steel Making Process', *EJ*, 1967, 531–58.

27 A. Sen, *Employment, Technology and Development* (Oxford, 1975).

28 See, in the same vein, M. I. Nadiri, 'Some Approaches to the Theory and Measurement of Total Factor Productivity: A Survey', *JEL*, 1970, 1170–71.

29 J. de V. Graaff, *Theoretical Welfare Economics* (Cambridge, 1963) is still an excellent introduction to the theory of welfare. See also E. J. Mishan, 'A Survey of Welfare Economics, 1939–1959', *EJ*, 1960, 197–256, reprinted in *Welfare Economics, Ten Introductory Essays*, 2nd edn (New York, 1969), pp. 1–86; M. Dobb, *Welfare Economics and the Economics of Socialism* (Cambridge, 1969), pp. 3–118; D. M. Winch, *Analytical Welfare Economics* (Harmondsworth, 1971).

30 See N. Kaldor, 'Welfare Propositions and Interpersonal Comparisons of Utility', *EJ*, 1939, 549–52; J. R. Hicks, 'The Valuation of Social Income', *Economica*, 1940, 105–24; T. Scitosky, 'A Note on Welfare Propositions in Economics',. *RES*, 1941, 77–88, reprinted in *Papers on Welfare and Growth* (London, 1964), pp. 123–38; *idem*, 'A Reconsideration of the Theory of Tariffs', *RES*, 1942, 89–110 (reprinted in the same collection, pp. 139–66); P. A. Samuelson, *Foundations of Economic Analysis*, 3rd edn (Cambridge, Mass., 1953), pp. 243–53; W. J. Baumol, *Welfare Economics and the Theory of the State*, 2nd edn (London, 1965), pp. 161–72.

31 J. R. Hicks, *Capital and Growth* (London, 1965), pp. 202–3.

32 P. Hennipman, 'Tweërlei interpersonele nutsvergelijkingen' ('Two Kinds of Interpersonal Comparison of Utility'), in *Mens en Keuze (Man and Choice)*, a collection of essays dedicated to Professor S. Korteweg (Amsterdam, 1972), p. 84.

33 A. C. Pigou, *Economics of Welfare*, 4th edn (London, 1932), pp. 188–90.

34 Pigou made this distinction as early as 1912 in his book *Wealth and Welfare* (London, 1912), p. 162.

35 F. Machlup, 'Erfindung und technische Forschung', *Handwörterbuch der Sozialwissenschaften*, vol. 3 (Tübingen, 1961), pp. 280–9.

36 J. Schmookler, *Invention and Economic Growth* (Cambridge, Mass., 1966), pp. 57ff.; *idem*, 'Changes in Industry and in the State of Knowledge as Determinants of Industrial Invention', in *The Rate and Direction of Inventive Activity: Economic and Social Factors*, ed. R. R. Nelson (Princeton, 1962), pp. 195ff.

37 R. R. Nelson, 'The Link between Science and Invention: The Case of the Transistor', in *The Rate and Direction of Inventive Activity*, pp. 548ff.

38 J. Jewkes, 'How much Science?', *EJ*, 1960, 1–6.

39 N. Rosenberg, 'Science, Invention and Economic Growth', *EJ*, 1974, 106.

40 K. J. Arrow, 'Economic Welfare and the Allocation of Resources for Invention', in *The Rate and Direction of Inventive Activity*, pp. 609–25,

also reprinted in *Economics of Information and Knowledge*, ed. D. M. Lamberton (Harmondsworth, 1971), pp. 141–59 and in *The Economics of Technological Change*, ed. N. Rosenberg (Harmondsworth, 1971), pp. 164–81. An important collection of studies is K. J. Arrow, *Essays in the Theory of Risk-Bearing* (Amsterdam, 1971).

41 On increasing returns to use, see also F. Machlup, 'The Supply of Inventors and Inventions', *Weltwirtschaftliches Archiv*, 1960, 210–54, especially pp. 236ff., also included in the collection of essays ed. R. R. Nelson, op. cit. See also F. Machlup, *The Production and Distribution of Knowledge in the United States* (Princeton, 1962).

42 H. Demsetz, 'Information and Efficiency: Another Viewpoint', *Journal of Law and Economics*, 1969, 1–22, reprinted in Lamberton's collection of essays, op. cit., pp. 160–86. See also B. S. Yamey, 'Monopoly, Competition and the Incentive to Invent: A Comment', *The Journal of Law and Economics*, 1970, 253–6; M. I. Kamien and N. L. Schwartz, 'Market Structure, Elasticity of Demand and Incentive to Invent', *The Journal of Law and Economics*, 1970, 241–52. Cf. also a summary of the discussion in C. K. Rowley, *Antitrust and Economic Efficiency* (London, 1973), pp. 33–45.

43 C. Aislabie, 'The Economic Efficiency of Information Producing Activities', *The Economic Record*, 1972, 575–83. See also J. Hirschleifer, 'Where Are We in the Theory of Information?', *The American Economic Review*, 1973, 31–9.

44 K. J. Arrow, 'Limited Knowledge and Economic Analysis', *AER*, 1974, 1–10.

45 Cf. D. Usher, 'The Welfare Economics of Invention', *Economica*, 1964, 279–87, where, incidentally, an insufficient distinction is made between inventions and innovations.

46 Cf. also T. Scitovsky and A. A. Scitovsky, 'What Price Economic Progress?', *Yale Review*, 1959, reprinted in *Papers on Welfare and Growth*, ed. T. Scitovsky (London, 1964), pp. 209ff.

47 For this see also B. S. Frey, 'Interactions Between Preferences and Consumption in Economic Development', *Scottish Journal of Political Economy*, 1973, 53–63.

48 See, further, E. J. Mishan's publications, especially *The Costs of Economic Growth* (London, 1967).

49 See M. Blaug, *An Introduction to the Economics of Education* (London, 1971), pp. 61–136; also A. Razin, 'Optimum Investment in Human Capital', *RES*, 1972, 455–60, where the formal structure of an optimization problem is chosen explicitly. See also *Education and Economic Development*, ed. C. A. Anderson and B. J. Bouman (Chicago, 1965). The screening hypothesis has been discussed by R. Layard and G. Psacharopoulos in 'The Screening Hypothesis and Returns to Education', *JPE*, 1974, 985–98. It is assumed in this hypothesis that the differences in earning between people educated to different levels ('inter-educational earning differentials') show no direct productivity-increasing effect of education, but merely reflect pre-existing differences in ability.

50　Cf. also A. Silberston, 'The Patent System', *Lloyds Bank Review*, 1967, 32–44, reprinted in *Economics of Information and Knowledge*, ed. D. M. Lamberton, pp. 224–38. See, in particular, the important work by C. T. Taylor and A. Silberston, *The Economic Impact of the Patent System* (London, 1973).

51　See, J. de V. Graaff, *Theoretical Welfare Economics* (Cambridge, 1963); E. J. Mishan, *Welfare Economics, Ten Introductory Essays*, 2nd edn (New York, 1969); idem, *The Costs of Economic Growth* and *Growth: The Price we Pay* (London, 1969); D. M. Winch, *Analytical Welfare Economics* (London, 1971); T. Scitovsky, *Welfare and Competition*, 2nd edn (London, 1971); E. J. Mishan, 'The Postwar Literature on Externalities: An Interpretative Essay', *JEL*, 1971, 1–28. On the environment, see especially A. M. Freeman and A. V. Kneese, *The Economics of Environmental Policy* (New York, 1973); R. S. Scorer, *Pollution in the Air* (London, 1973); P. A. Victor, *Economics of Pollution* (London, 1972); idem, *Pollution: Economy and Environment* (London, 1972).

52　For this see also J. R. Hicks, 'The Mainspring of Economic Growth', *The Swedish Journal of Economics*, 1973, 336–48.

53　A pioneering work is F. P. Ramsey, 'A Mathematical Theory of Saving', *EJ*, 1928, 543ff.; see also a review article by A. Heertje, 'On the Optimum Rate of Savings', *Weltwirtschafliches Archiv*, 1963, 7–43, and the literature mentioned there; R. Bharadwaj and T. Mansharan, 'On the Theory of Optimal Savings with Finite Planning Horizon', *SEJ*, 1966/67, 264–7 and S. Chakravarty, *Capital and Development Planning* (Cambridge, Mass., 1969). A good survey of developments after 1962 is B. S. Frey, 'Optimales Wachstum, Übersicht und Kritik', *Jahrbücher für Nationalökonomie und Statistik*, 1970, 9–30; see also J. Schuman, 'Zur Theorie optimalen wirtschaflichen Wachstums', *Zeitschrift für die gesamte Staatswissenschaft*, 1969, 1–16.

54　A. Bergson, 'A Reformation of Certain Aspects of Welfare Economics', *QJE*, 1938, 310–34, reprinted in *Readings in Welfare Economics*, ed. K. J. Arrow and T. Scitovsky (London, 1969), pp. 7–25 and in A. Bergson, *Essays in Normative Economics* (Cambridge, Mass., 1966), pp. 3–26.

55　J. Tinbergen and H. C. Bos, *Mathematical Models of Economic Growth* (New York, 1962), p. 24. A Recent illustration is to be found in E. Asher, 'Consumers' Time-Preference and Welfare Economics under Socialism', *Scottish Journal of Political Economy*, 1973, 283–90.

56　D. H. Robertson, *Lectures on Economic Principles*, vol. 2 (London, 1958), p. 89.

57　See, e.g., M. Morishima, *Theory of Economic Growth* (Oxford, 1969), pp. 213–25; E. Burmeister and A. R. Dobell, *Mathematical Theories of Economic Growth* (London, 1970), pp. 352–427; see also Wan, *Economic Growth*, pp. 267ff.

58　Kurz has developed a model in which the utility is also dependent on the stock of capital goods per head, M. Kurz, 'Optimal Economic Growth and Wealth Effects', *IER*, 1968, 348–57.

59 T. C. Koopmans, 'On the Concept of Optimal Economic Growth', in *Studyweek on the Econometric Approach to Planning* (Chicago, 1965), pp. 224–87, reprinted in *Scientific Papers of Tjalling Koopmans* (Berlin, 1970), pp. 485–547; another important study is M. Inagaki, 'Optimal Economic Growth', in *Shifting Finite Versus Infinite Time Horizon* (Amsterdam, 1970); T. C. Koopmans, 'Concepts of Optimality and their Uses', *forthcoming Nobel Prize Lecture*.

60 Cf., e.g., T. C. Koopmans, 'Objectives, Constraints and Outcomes in Optimal Growth Models', *Econometrica*, 1967, 1–15; M. Boyer, 'An Optimal Growth Model with Stationary Non-Additive Utilities', *CJEPS*, 1975, 216–37.

61 A. Asimakopoulos, 'Optimal Economic Growth and Distribution and the Social Utility Function', *CJEPS*, 1968, 541.

62 P. A. Samuelson, 'Turnpike Theorems Even Though Tastes Are Intertemporally Dependent', *Western Economic Journal*, 1971, 21–6.

63 H. E. Ryder Jr. and G. M. Heal, 'Optimum Growth with Intertemporally Dependent Preferences', *RES*, 1973, 1–31.

64 See C. C. von Weizsäcker, 'Existence of Optimal Programs of Accumulation for an Infinite Time Horizon', *RES*, 1965, 85–104.

65 See R. Radner, 'Paths of Economic Growth that are Optimal with Regard Only to Final States: a Turnpike Theorem', *RES*, 1961, 98–104; H. Nikaido, 'Persistence of Continual Growth near the von Neumann Ray: A Strong Version of the Radner Turnpike Theorem', *Econometrica*, 1964, 151–62; D. Cass, 'Optimum Growth in an Aggregative Model of Capital Accumulation: A Turnpike Theorem', *Econometrica*, 1966, 833–50; a relatively simple account of these complicated problems has been given by J. Vanek in *Maximal Economic Growth* (New York, 1968).

66 B. Peleg and H. E. Ryder Jr., 'On Optimal Consumption Plans in a Multi-Sector Economy', *RES*, 1972, 159–69. See also Mohamed El-Hodiri, 'The Ramsey Problem in Optimal Economic Growth under Resource Limitation', *Metroeconomica*, 1971, 197–219; B. Peleg and H. E. Ryder Jr., 'The Modified Golden Rule of a Multi-Sector Economy', *Journal of Mathematical Economics*, 1974, 192–8.

67 Horvat calls this approach, not altogether without reason, 'operationally meaningless', B. Horvat, 'A Model of Maximal Economic Growth', *Kyklos*, 1972, 215.

68 See J. R. Hicks, *Capital and Growth*, p. 203.

69 B. A. Forster, 'Optimal Consumption Planning in a Polluted Environment, *The Economic Record*, 1973, 534–45.

70 K. Shell, 'Optimal Programs of Capital Accumulation for an Economy in which there is Exogenous Technical Change' in *Essays on the Theory of Optimal Economic Growth*, ed. K. Shell (Cambridge, Mass., 1967), pp. 1–30. A model of exogenous embodied technical change, based on Shell's analysis, has been developed by B. Goldar, 'Optimal Programs of Capital Accumulation Under Exogenous Embodied Technical Change', *JET*, 1974, 224–9.

71 E. S. Phelps, 'The Golden Rule of Accumulation: A Fable for

Growthmen', *AER*, 1961, 638–43; *idem, Golden Rules of Economic Growth* (Amsterdam, 1967); C. C. von Weizsäcker, *Wachstum, Zins und optimale Investitionsquote* (Basle, 1962); J. Robinson, 'A Neoclassical Theorem', *RES*, 1962, 219–26. A variable rate of population growth is chosen as the starting point in R. C. Merton, 'A Golden Rule for Welfare-Maximization in an Economy with a Varying Population Growth Rate', *Western Economic Journal*, 1969, 307–18. An endogenous population growth is also found in a model of R. Sato and E G. Davies, 'Optimal Savings Policy when Labor Grows Endogenously', *Econometrica*, 1971, 877–97.

72   For the differences between the Ramsay-type models of optimal growth and the Golden Rule, see E. S. Phelps, op. cit., pp. 69–103.

73   J. A. Mirrlees, 'Optimum Growth when Technology is Changing', *RES*, 1967, 95–124; B. Goldar, 'Optimal Programs of Capital Accumulation under Exogenous Embodied Technical Change', *JET*, 1974, 224–9.

74   E. Sheshinski, 'Optimal Accumulation with Learning by Doing' in *Essays on the Theory of Optimal Economic Growth*, ed. K. Shell, pp. 31–52.

75   E. S. Phelps, *Golden Rules of Economic Growth*, pp. 129–33.

76   The danger of confusion in this respect has already been indicated by M. Allais in a brilliant book, *Economie et Intérêt* (Paris, 1947), pp. 550–1.

77   This appears from C. Duncan MacRae, 'Equilibrium, Efficiency and the Golden Rule', *QJE*, 1974, 143–8.

78   D. M. Nuti, 'Capitalism, Socialism and Steady Growth', *EJ*, 33–57. Nuti rightly ignores the possibility of a wage-interest frontier with a positive gradient. See also G. O. Orosel, 'A Note on the Factor-Price Frontier', *Zeitschrift für Nationalökonomie*, 1973, 103–14; D. M. Nuti, 'The Wage-Interest Frontier: A Reply to Dr Orosel', *Zeitschrift für Nationökonomie*, 1975, 177–86; and G. O. Orosel, 'Shut-Down Costs, Borrowing from Nature and Negative Capital: A Rejoinder to Professor Nuti', *Zeitschrift für Nationalökonomie*, 1975, 187–94. For the 'truncation problem' in this context see D. M. Nuti, 'On the Truncation of Production Flows', *Kyklos*, 1973, 485–96.

79   See A. Heertje, 'The Cambridge Controversy and Welfare' in *Policy and Precision, Essays in Honour of Professor P. de Wolff*, ed. J. S. Cramer, A. Heertje and P. E. Venekamp (Amsterdam, 1976).

80   See K. Navis and J. Vaizey, *The Economics of Research and Technology* (London, 1973), p. 120.

81   The same view has been put forward by J. Hirschleifer in 'The Private and Social Value of Information and the Reward to Inventive Activity', *AER*, 1971, 573.

82   See W. J. Fellner, 'Trends in the Activities Generating Technological Progress', *AER*, 1970, 1.

83   See J. Boswell, *The Rise and Decline of Small Firms* (London, 1973), p. 197.

84   Cf. S. A. Hetzler, *Applied Measures for Promoting Technological Growth* (London, 1973), pp. 38ff.

85   E. F. Schumacher, *Small is beautiful* (London, 1973), pp. 136–49.

86 See, e.g., E. J. Mishan, *Cost-Benefit Analysis* (London, 1971); *idem*, *Elements of Cost-Benefit Analysis* (London, 1972); A. K. Dasgupta and D. W. Pearce, *Cost-Benefit Analysis:Theory and Practice* (London, 1972); a simple introduction to this subject is given in D. W. Pearce, *Cost-Benefit Analysis* (London, 1971).

87 The same view is expressed by T. Wilson in 'The Price of Growth', *EJ*, 1963, 617; cf. also S. Brittan, *Is there an Economic Consensus?* (London, 1973), which discusses the way economists arrive at an opinion.

88 D. L. Meadows, *The Limits to Growth* (New York, 1972). See also the special issue of *The Swedish Journal of Economics*, 1971, 1–156, with contributions from several authors on environmental economics.

89 A broadly based and rigorous criticism is to be found in *Thinking About the Future, A Critique of the Limits to Growth*, ed. J. S. Cole, C. Freeman, M. Ja Hoda and K. L. Pavitt (London, 1973). Another study of importance is W. D. Nordhaus, 'World Dynamics: Measurement without Data', *EJ*, 1973, 1156–83.

90 See W. Beckerman, 'Economists, Scientists and Environmental Catastrophe', *OEP*, 1971, 327–44. Beckerman has been criticized by L. S. Brown *et al.*, 'Are There Real Limits to Growth? A Reply to Beckerman', *OEP*, 1973, 455–60.

91 E. J. Mishan, 'Review of W. Beckerman's "In Defence of Economic Growth" ', *JPE*, 1975, 873.

92 W. Beckerman, *In Defence of Economic Growth* (London, 1974), chapters 5 and 6; A. A. Walters, *Noise and Prices* (Oxford, 1975).

93 R. M. Solow, 'The Economics of Resources and the Resources of Economics', *AER*, Papers and Proceedings, 1974, 13; see also J. E. Meade, *The Theory of Indicative Planning* (Manchester, 1975) and the special issue of *The Review of Economic Studies*, 1974, 1–152, on the economics of exhaustible resources.

94 See, e.g., R. S. Scorer, *Pollution in the Air* (London, 1975).

95 See J. S. Bain, *Environmental Decay* (Boston, 1973), p. 230; see also *Managing the Environment*, ed. A. V. Kneese *et al.* (London, 1971).

96 E. J. Mishan, 'Growth and Antigrowth: What are the Issues?', in *The Economic Growth Controversy*, ed. A. Weintraub (London, 1974), pp. 3–38.

97 M. Kranzberg, 'Can Technological Progress Continue to Provide for the Future?', in *The Economic Growth Controversy*, pp. 62–81.

98 W. S. Jevons, *The Coal Question* (London, 1866).

99 N. Georgescu-Roegen, *The Entropy Law and the Economic Process* (Cambridge, Mass., 1971), pp. 293ff.See also K. E. Boulding, *Economics as a Science* (New York, 1970), pp. 44ff.

100 See R. C. d'Arge and K. C. Kogiku, 'Economic Growth and the Environment', *RES*, 1973, 76. See also J. Tinbergen, 'Exhaustion and Technological Development: A Macro-Dynamic Policy Model', *Zeitschrift für Nationalökonomie*, 1973, 213–34.

CHAPTER 13: THE SHIFTING FRONTIERS

1 J. Schmookler, *Patents, Invention and Economic Change*, ed. Z.

Griliches and L. Hurwicz (Cambridge, Mass., 1972), p. 84.

2 T. Veblen, *The Theory of Business Enterprise* (New York, 1904). For a comparison of his views with Schumpeter's see L. A. O'Donnell, 'Rationalism, Capitalism and the Entrepreneur: the Views of Veblen and Schumpeter', *HPE*, 1973, 199–214. Veblen's views have been compared with those of Galbraith in G. C. Leathers and J. S. Evans, 'Thorstein Veblen and the New Industrial State', *HPE*, 1973, 420–37.

3 H. Gintis, 'Neo-Classical Welfare Economics and Individual Development', *QJE*, 1972, 572–99; *idem*, 'Welfare Criteria with Endogenous Preferences: the Economics of Education', *IER*, 1974, 415–30; see also T. Parson, 'Welfare Economics and Individual Development', *QJE*, 1975, 280–90; H. Gintis' comment, ibid., 291–302; and D. K. Foley, 'Problem vs Conflicts: Economic Theory and Ideology', *AER*, 1975, 231–6.

4 See A. G. Gruchy, 'Contemporary Economic Thought' in *The Contribution of Neo–Institutional Economics* (London, 1972), p. 296.

5 See M. Abramowitz and P. A. David, 'Reinterpreting Economic Growth: Parables and Realities', *AER*, 1973, 429; see also N. Rosenberg, 'Problems in the Economist's Conceptualization of Technological Innovation', *HPE*, 1975, 456–81.

6 See N. Rosenberg, 'On Technological Expectations', *EJ*, 1976, 523–35 and *Perspectives on Technology* (London, 1976).

7 J. M. Keynes, *The General Theory of Employment, Interest and Money* (London, 1936), pp. 217–21. See also L. Vandone, 'Institutional, Structural and Technological Changes in Decentralized Decision Systems and Economic Stagnation', *Rivista di Politica Economica, Selected Papers*, 1973, 186–242.

8 H. Kahn and A. J. Wiener, *The Year 2000, A Framework for Speculation on the Next Thirty-Three Years* (New York, 1967), p. 94

9 J. M. Keynes, *Essays in Persuasion* (London, 1931), p. 366.

10 See E. Jantsch, *Technological Planning and Social Futures* (London, 1972); *idem*, in *Technological Forecasting in Perspective* (Paris, 1967). L. Thiriet, 'Note sur une nouvelle méthode de prévision technologique', *Revue économique*, 1973, 316–30. The standard work on this subject is undoubtedly *A Guide to Practical Technological Forecasting*, ed. J. R. Bright and M. E. Schoeman (Englewood Cliffs, New Jersey, 1973). See also H. Bloem and K. Steinbruch, *Technological Forecasting in Practice* (Westmead, 1973); C. Robinson, *The Technology of Forecasting and the Forecasting of Technology* (London, 1972); and M. I. Kamien and N. L. Schwartz, 'Some Economic Consequences of Anticipating Technical Advance', *Western Economic Journal*, 1972, 123–38.

Index of Names

General Index

Abramowitz residual, 199, 200, 201, 202
abstract technology, 142
accumulation, Golden Rule of, 253, 255
administered prices, 228–9
aggregate production function, 131, 139, 156–7, 161, 167
agriculture
  investment capital, 8–9
  labour migration, 92, 203
  productivity, 9, 10
allocation of resources, 220–1, 242, 243, 245–8
ancient technology, 62–4
Austrian school, 83–4, 104, 128–9
automation, 108–9, 112–14, 187

business cycles, 101, 189–90, 203

Cambridge controversy, 160–1
capital
  accumulation, 11–12, 13, 17, 22, 41–2, 46, 47–9, 50–3, 54–61, 82, 185, 204
  augmentation, 82, 179–80, 184–5, 186, 213
  circulating, 85
  constant, 19–20, 41
  elasticity of, 197, 200
  fixed/circulating ratio, 14–17, 19
  formation, 43, 59, 83, 84, 97, 186, 187, 196, 252
  growth of, 184
  investment, 83–4
  marginal productivity of, 88, 162
  organic composition of, 41, 43, 45–6, 50–3, 54, 57, 60, 152, 153
  variable, 19–20, 41
capital goods, 48, 148–50, 154–5, 159, 168, 193–4, 248
  aggregation, 148–9
  new, 193, 199, 204, 249, 256, 258
  technical change embodied in, 173, 174–7, 182, 184, 187, 194, 199, 200, 204, 256, 258
capital/labour ratio, 152, 153

capital-saving techniques and inventions, 8, 10, 56–7, 162, 166, 169, 178–9, 180, 187, 244
capitalism, 59–60, 95, 99–100, 101, 104–5, 189, 208, 256
capitalistic production, 40–2
cartels, 229
centralization, 42
classical economics, 2, 7–37, 55–6, 60
Cobb–Douglas production function, 88, 93, 125–7, 130, 135, 141, 163, 168, 174–5, 184–5, 186, 193, 194, 195, 196, 197, 200, 201, 204
collective bargaining, 228
communications, 76, 92, 110–11, 119
compensation theory 25–34, 36–7, 42–3, 147–8, 185–8, 242, 244
competition, 42, 59, 207
  innovative, 208, 230
  non-price, 226
  perfect, 49, 121, 185, 209, 216
  policy on, 239
  price, 208
  vertical, 208
computers, 108, 110–12, 116
concentration
  conglomerate, 210, 211
  curve, 210
  horizontal, 210, 211, 220, 231
  vertical, 210, 211
conglomerate concentration, 210, 211
constant capital, 41
constant elasticity of substitution (CES), 127, 130, 135, 141, 193, 197, 198, 201
consumer goods, 247, 248, 255, 258
consumer preference, 96, 247–8, 255, 268
consumption functions, 191
convergence theory, 238–9
cost-benefit analysis, 264
cost controversy, 121
cost functions, 121–2
  marginal, 80
  total, 80
craftsmen, society based on, 40